高等教育应用型重点专业规划教材

New English Course in Engineering

新编工程英语

丛书主编：冯光华
主编：王 莹　周 丹　严红烨
副主编：李 萍

天津大学出版社
TIANJIN UNIVERSITY PRESS

图书在版编目（CIP）数据

新编工程英语 / 王莹，周丹，严红烨主编. 一天津：天津大学出版社，2018.5

高等教育应用型重点专业规划教材 / 冯光华主编

ISBN 978-7-5618-6129-5

Ⅰ.①新… Ⅱ.①王… ②周… ③严… Ⅲ.①工程技术－英语－高等学校－教材 Ⅳ.①TB

中国版本图书馆CIP数据核字（2018）第103944号

出版发行	天津大学出版社
地　　址	天津市卫津路92号天津大学内（邮编：300072）
电　　话	发行部：022-27403647
网　　址	publish.tju.edu.cn
印　　刷	天津泰宇印务有限公司
经　　销	全国各地新华书店
开　　本	185mm×260mm
印　　张	8.25
字　　数	230千
版　　次	2018年5月第1版
印　　次	2018年5月第1次
定　　价	26.00元

凡购本书，如有缺页、倒页、脱页等质量问题，烦请向我社发行部门联系调换

版权所有　侵权必究

丛书编委会

总主编：冯光华

编　委：（排名不分先后）

吴海燕	田　娟	房明星	陈　斯	胡　杨	严红烨
蔡丽慧	王电兴	陈　招	王　莹	李璐瑶	左　帆
李　游	杨永华	罗　莉	李　杨	张馨引	李丽花
张　慧	黄亚楠	王　晓	潘婷婷	陈　霞	李　奕
陈　霞	芦　佳	蔡　喆	黄　莉	王华英	胡　玲
孙川慧	张培芳	管春梅	刘　丽	薛海琴	贾　文
李　萍	关荆晶	何丽荣	毛兆婷	段友国	周　丹
石立国					

《新编工程英语》编委会

主　编：王　莹　周　丹　严红烨
副主编：李　萍
编　者：潘婷婷　胡　杨　杨永华　黄　莉

前言

　　为了更好地培养适应社会需要的应用技术型复合人才，《新编工程英语》根据编者的教学实践以及工程行业专家的实操经验，遵循"实用为主，够用为度，以应用为目的"的编写原则，紧密结合工程专业职业需求，立足专业前沿，与生产实际紧密结合，以期培养和提高学生实际运用工程英语的语言能力。

　　全书共分八个实用主题单元，依次为"土木工程专业和课程""设计过程""建筑材料""桥梁""建筑物类型""工程造价""标书和合同"和"施工"。与此同时，单元练习多样化，包含专业词汇（Special Terms）、实用句型（Sentence Patterns）、补全对话、阅读理解和文段翻译五个部分。这些主题单元的设置和练习的编排都紧紧围绕工程专业前沿，以帮助学生掌握英语技能和专业知识，提高学生工程专业英语阅读和翻译能力，培养专业职业素养为目的。《新编工程英语》集职业性、实用性、适时性和趣味性于一体；内容新颖，语言通俗易懂，图文并茂突出了形式上的多样性和直观性。本书最后还附录了各主题单元练习答案以及课文译文，方便学生自学与自查。

　　《新编工程英语》可作为应用技术型高校工程专业教材供应用技术型高校的师生们使用。

Unit 1 Civil Engineering 1

Unit 2 Design Process 17

Unit 3 Structural Materials 28

Unit 4 Bridges 41

Unit 5 Building Code 55

Unit 6 Construction Cost Estimate 64

Unit 7 Tender Documents and Contracts 78

Unit 8 Construction 87

练习答案及参考译文 97

Unit 1　Civil Engineering

Learning Objectives

After completing this unit, you will be able to do the following:
- √ Grasp the main idea and the structure of the text;
- √ Master the key language points and grammatical structure in the text;
- √ Understand the compulsory and basic courses;
- √ Conduct a series of reading, listening, speaking and writing activities related to the theme of the unit.

Outline

The following are the main sections in this unit.
1. Warm-up Activity
2. Text
3. Words and Expressions
4. Exercises

Terms of Civil Engineering

In this unit, you will learn the meanings of terms listed below.
computer aided design
photogrammetry
pipeline
structural engineering
mapping

Vocabulary

Listed below are some words appearing in this unit that you should make them a part of your vocabulary.
irrigation
drainage
slurry
municipal
excavate
planner

Looking Ahead

Civil engineers deal with the design and construction of roads, buildings, airports, tunnels, dams, bridges and water supply or sewage systems. This course enables you to progress and qualify as a chartered civil or structural engineer, and provides a strong base of high-quality technical abilities together with good management and personal skills.

Introduction

Civil engineering is the application of physical and scientific principles for solving the problems of society, and its history is intricately linked to advances in understanding of physics and mathematics throughout history. Because civil engineering is a wide ranging profession, including several separate specialized subdisciplines, its history is linked to knowledge of structures, materials science, geography, geology, soils, hydrology, environment, mechanics and other fields.

◈ Warm-up Activity

What do you think of careers in civil engineering?

📋 Text

Civil Engineering

Civil engineering is the planning, design, construction and management of the built environment. This environment includes all structures built according to scientific principle from irrigation and drainage systems to rocket-launching facilities.

Civil engineers build roads, bridges, tunnels, dams, harbors, power plants, water and sewage systems, hospitals, schools, mass transit, and other public facilities essential to modern society and large population concentrations. They also build privately owned facilities such as airports, railroads, pipelines, skyscrapers, and other large structures designed for industrial, commercial or residential use. In addition, civil engineers plan, design and build complete cities and towns, and more recently have been planning and designing space platforms to house self-contain communities.

The word *civil* derives from the Latin for *citizen*. In 1782, Englishman John Smeaton used the term to differentiate his nonmilitary engineering work from that of the military engineers who predominated at the time. Since then, the term of civil engineering has often been used to refer to engineers who build public facilities, although the field is much broader.

Scope

Because it is so broad, civil engineering is subdivided into a number of technical specialties. Depending on the type of project, the skills of many kinds of civil engineer specialists may be needed. When a project begins, the site is surveyed and mapped by civil engineers who experiment to determine if the earth can bear the weight of the project. Environmental specialists study the project's impact on the local area, the potential for air and groundwater pollution, the project's impact on local animal and plant life, and how the project can be designed to meet government requirements aimed at protecting the environment. Transportation specialists determine what kind of facilities are needed to ease the burden on local roads and other transportation networks that will result from the completed project. Meanwhile, structural specialists use preliminary data to make detailed designs, plans and specifications for the project. Supervising and coordinating the work of these civil engineer specialists, from beginning to end of the project, are the construction management specialists. Based on information supplied by the other specialists, construction management civil engineers estimate quantities and costs of materials and labor, schedule all work, order materials and equipments for the job, hire contractors and subcontractors, and perform other supervisory work to ensure the project is completed on time and as specified.

Throughout any given project, civil engineers make extensive use of computers. Computers are used to design the project's various elements (computer aided design, CAD) and to manage it. Computers are a necessity for the modern civil engineer because they permit the engineer to efficiently handle the large quantities of data needed in determining the best way to construct a project.

Structure engineering In this specialty, civil engineers plan and design structures of all types, including bridges, dams, power plants, supports for equipment, special structures for offshore projects, the United States space program, transmission towers, giant astronomical and radio telescopes, and many other kinds of projects. Using computers, structural engineers determine the forces a structure must resist: its own weight, wind and hurricane forces, temperature changes that expand or contract construction materials, and earthquakes. They also determine the combination of appropriate materials: steel, concrete, plastic, stone, asphalt, brick, aluminum, or other construction materials.

Water resources engineering Civil engineers in this specialty deal with all aspects of the physical control of water. Their projects help prevent floods, supply water for cities and for irrigation, manage and control rivers and water runoff, and maintain beaches and other waterfront facilities. In addition, they design and maintain harbors, canals, and locks, build huge hydroelectric dams and smaller dams and water impoundments of all kinds, help to design offshore structures, and determine the location of structures affecting navigation.

Geotechnical engineering   Civil engineers who specialize in this field analyze the properties of soils and rocks that support structures and affect structural behavior. They evaluate and work to minimize the potential settlement of buildings and other structures that stems from the pressure of their weight on the earth. These engineers also evaluate and determine how to strengthen the stability of slopes and fills and how to protect structures against earthquakes and the effects of groundwater.

Pipeline engineering   In this branch of civil engineering, engineers build pipelines and related facilities which transport liquids, gases, or solids ranging from coal slurries (mixed coal and water) and semi liquid wastes, to water, oil, and various types of highly combustible and noncombustible gases. The engineers determine pipeline design, the economic and environmental impact of a project on regions it must traverse, the type of materials to be used— steel, concrete, plastic, or combinations of various materials, installation techniques, methods for testing pipeline strength, and controls for maintaining proper pressure and rate of flow of materials being transported. When hazardous materials are being carried, safety is a major consideration as well.

Environmental engineering   In this branch of engineering, civil engineers design, build, and supervise systems to provide safe drinking water and to prevent and control pollution of water supplies, both on the surface and underground. They also design, build, and supervise projects to control or eliminate pollution of the land and air. These engineers build water and waste water treatment plants, and design air scrubbers and other devices to minimize or eliminate air pollution caused by industrial processes, incineration, or other smoke-producing activities. They also work to control toxic and hazardous wastes through the construction of special dump sites or the neutralizing of toxic and hazardous substances. In addition, the engineers design and manage sanitary landfills to prevent pollution of surrounding land.

Transportation engineering   Civil engineers working in this specialty build facilities to ensure safe and efficient movement of both people and goods. They specialize in designing and maintaining all types of transportation facilities, highways and streets, mass transit systems, railroads and airfields, ports and harbors. Transportation engineers apply technological knowledge as well as consideration of the economic, political, and social factors in designing each project. They work closely with urban planners, since the quality of the community is directly related to the quality of the transportation system.

Construction engineering Civil engineers in this field oversee the construction of a project from the beginning to the end. Sometimes called project engineers, they apply both technical and managerial skills, including knowledge of construction methods, planning, organizing, financing, and operating construction projects. They coordinate the activities of virtually everyone engaged in the work: the surveyors; workers who lay out and construct the temporary roads and ramps, excavate for the foundation, build the forms and pour the concrete; and workers who build the steel framework. These engineers also make regular progress reports to the owners of the structure.

Community and urban planning Those engaged in this area of civil engineering may plan and develop community within a city, or entire cities. Such planning involves far more than engineering consideration; environmental, social, and economic factors in the use and development of land and natural resources are also key elements. These civil engineers coordinate planning of public works along with private development. They evaluate the kinds of facilities needed, including streets and highways, public transportation systems, airports, port facilities, water-supply and waste water-disposal systems, public buildings, parks, and recreational and other facilities to ensure social and economic as well as environmental well-being.

Photogrammetry, surveying, and mapping The civil engineers in this specialty precisely measure the Earth's surface to obtain reliable information for locating and designing engineering projects. This practice often involves high-technology methods such as satellite and aerial surveying, and computer processing of photographic imagery. Radio signal from satellites, scans by laser and sonic beams, are converted to maps to provide far more accurate measurements for boring tunnels, building highways and dams, plotting flood control and irrigation project, locating subsurface geologic formations that may affect a construction project, and a host of other building uses.

Other specialties

Two additional civil engineering specialties that are not entirely within the scope of civil engineering but are essential to the discipline are engineering management and engineering teaching.

Engineering management Many civil engineers choose careers that eventually lead to management. Others are able to start their careers in management positions. The civil engineer-manager combines technical knowledge with an ability to organize and coordinate worker power, materials, machinery, and money. These engineers may work in government — municipal, county, state, or federal; in the U.S. Army Corps of Engineers as military or civilian management engineers; or in semiautonomous regional or city authorities or similar organizations. They may also manage private engineering firms ranging in size from a few employees to hundreds.

Engineering teaching   The civil engineer who chooses a teaching career usually teaches both graduate and undergraduate students in technical specialties. Many teaching civil engineers engage in basic research that eventually leads to technical innovations in construction materials and methods. Many also serve as consultants on engineering projects, or on technical boards and commissions associated with major projects.

Words and Expressions

planning	['plænɪŋ]	n. 计（规）划
irrigation	[ˌɪrɪ'geɪʃn]	n. 灌溉
drainage	['dreɪnɪdʒ]	n. 排水
launch	[lɔ:ntʃ]	vt. 发射
sewage	['su:ɪdʒ]	n. 污水，（下水道里的）污物
pipeline	['paɪplaɪn]	n. 管道
skyscraper	['skaɪskreɪpə(r)]	n. 摩天大楼
residential	[ˌrezɪ'denʃl]	adj. 住宅的
predominate	[prɪ'dɒmɪneɪt]	vt. 支配，在……中占优势
specialty	['speʃəltɪ]	n. 专业
placement	['pleɪsmənt]	n. 放置、安置
sewer	['su:ə(r)]	n. 污水管
geotechnical	[ˌdʒi:əu'teknɪkəl]	adj. 土工技术的，岩土工程技术的
specification	[ˌspesɪfɪ'keɪʃn]	n. 规格，说明书
contractor	[kən'træktə]	n. 承包人
subcontractor	[ˌsʌbkən'træktə(r)]	n. 转包商
supervisory	['sju:pəˌvaɪzərɪ]	adj. 管理（监督）的
computer aided design		计算机辅助设计
structural engineering		结构工程
offshore	[ˌɒf'ʃɔ:]	adj. 近海的

astronomical	[ˌæstrə'nɒmɪk(ə)l]	adj. 天文（学）的
hurricane	['hʌrɪkən]	n. 飓风
asphalt	['æsfælt]	n. 沥青
runoff	['rʌnɒf]	n. 径流，流走的东西
lock	[lɒk]	n. 水闸
impoundment	[ɪm'paʊndmənt]	n. 蓄水
settlement	['se(ə)tlm(ə)nt]	n. 沉淀
scrubber	['skrʌbə]	n. 洗涤器
incineration	[ɪnˌsɪnə'reɪʃn]	n. 焚化
toxic	['tɒksɪk]	adj. 有毒的
hazardous	['hæzədəs]	adj. 有危险的
neutralize	['nju:trəlaɪz]	v. （使）中和
dump	[dʌmp]	n. 垃圾场
sanitary	['sænɪt(ə)rɪ]	adj. （环境）卫生的
airfield	['eəfi:ld]	n. （飞）机场
planner	['plænə]	n. 规划人员
slurry	['slʌrɪ]	n. 泥浆，残渣
combustible	[kəm'bʌstɪ(ə)bl]	adj. 易燃的
oversee	[ˌəʊvə'si:]	v. 监督，管理
managerial	[mænə'dʒi:rɪəl]	adj. 管理的
surveyor	[sə'veɪə]	n. 测量员
ramp	[ræmp]	n. 斜坡
excavate	['ekskəveɪt]	v. 挖掘
recreational	[ˌrekrɪ'eɪʃnl]	adj. 消遣的
well-being	['wel'bi:ɪŋ]	n. 福利，幸福
photogrammetry	[ˌfəʊtə(ʊ)'græmɪtrɪ]	n. 摄影测量法
surveying	[sə'veɪɪŋ]	n. 测量
mapping	['mæpɪŋ]	v. 绘图
aerial	['eərɪəl]	adj. 空中的
photographic	[ˌfəʊtə'græfɪk]	adj. 摄影的
imagery	['ɪmɪdʒ(ə)rɪ]	n. 成像
bore	[bɔ:]	v. 钻孔
geologic	[ˌdʒɪə'lɒdʒɪk]	adj. 地质的
municipal	[mjʊ:'nɪsɪp(ə)l]	adj. 市政的

semiautonomous	[ˌsemɪɔː'tɒnəməs]	*adj.* 半自治性的
innovation	[ˌɪnə'veɪʃn]	*n.* 革新
consultant	[kən'sʌltənt]	*n.* 顾问
be essential to		对……必要的
derive from		来源于
be used to		被用于
be subdivided into		再被细分
aim at		目的在于
stem from		产生（起源）于
work with		与……一道工作
be related to		与……有关
range from A to B		在 A 到 B 的范围内
along with		与……一道
a host of		许多
within the scope of		在……范围内
serve as		用作，充当

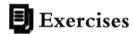

Exercises

Part One: Special Terms

1. differentiate A from B　　　_____
2. impact on　　　　　　　　　_____
3. meet the requirements of　　_____
4. 工程管理　　　　　　　　　_____
5. 基础研究　　　　　　　　　_____
6. 市政工程　　　　　　　　　_____

Part Two : Situational Conversation

(**Helen:** a visitor   **Xiaoli:** a student majoring in civil and industrial architecture)

Helen: You are a student, aren't you?

Xiaoli: Yes, but how did you know?

Helen: I saw it from your words and appearance, as you look very polite and elegant with a pair of glasses.

Xiaoli: You've good eyesight to know what I do.

Helen: Is it so? Can you tell me what major you are studying?

Xiaoli: Civil and industrial architecture.

Helen: How do you think of your major?

Xiaoli: Wonderful. I think I like it very much.

Helen: Why so, young man?

Xiaoli: As you know, all the buildings are set up by hard-working and bright builders from ancient times to the present. They are really great. So I determined to study architecture, and after graduation I am going to be a builder and to build up high buildings and large mansions with my own hands.

Helen: Your dream of building millions of apartments for the country and the people must be realized, young man. I know you enjoy your major, and I believe you must be a top student in your college, and hope you will be a builder of benefiting people.

Xiaoli: Thank you for your encouragement. I am sure to treasure the good chance and to study hard to realize my dream.

Helen: I'm glad to hear that. Can you tell me how many courses there are in your major?

Xiaoli: More than ten, I suppose. Construction material, architecture of houses, construction technique, construction organizations and mechanics, etc., are required courses; while English and political economics, etc., belong to basic courses.

Helen: So many! How do the teachers teach you?

Xiaoli: They seriously clarify the book knowledge from the shallower to the deeper and from the easier to the more advanced, and explain profound theories in simple languages.

Helen: I think it is the so-called programmed instruction. By the way, what about the teaching facilities?

Xiaoli: Very good. The basic theoretical knowledge of architecture is taught mainly in the classroom, while the operational skills are trained and practiced in the modernized architectural labs as well as on the cooperated construction worksites.

Helen: For vocational colleges, it is effective and practical to integrate theory with practice and to do practice geared to the needs of the jobs. Is your college large?

Xiaoli: Not very large.

Helen: Oh, I know. Would you please show me around your campus?

Xiaoli: It's my pleasure. This way, please!

Helen: Thank you very much.

Notes:

1. "…as you look very polite and elegant with a pair of glasses" 翻译为 "因为你戴着一副眼镜，看上去温文而雅。" 此句中的 as 表原因。

2. "Your dream of building millions of apartments for the country and the people must be realized." 翻译为 "你建成大厦千万栋，兴邦立国为人民的理想一定能实现。"

3. "required courses" 意为 "必修课"，"basic courses" 意为 "基础课"。

4. "…from the shallower to the deeper and from the easier to the more advanced" 的意思是 "由浅入深，由易到难"。

5. "…it is the so-called programmed instruction" 意思是 "这就是所谓的程序教学"。

6. "…the operational skills are trained and practiced in the modernized architectural labs as well as on the cooperated construction worksites" 意思是 "操作技能在现代化的建筑实验室以及拥有合作关系的建筑工地得以训练和实践"。

7. "…to integrate theory with practice" 意思为 "理论联系实际"。

8. "…to do practice geared to the needs of the jobs" 意思为 "对口实习"，"be geared to" 意思是 "适应……的需要，面向"。此处，"geared to…" 为过去分词短语作定语修饰 "practice"。

Exercise 1: Sentence Patterns

1. Can you tell me what major you are studying?
你能告诉我你现在学什么专业吗？

2. How do you think of /about your major?
你认为你的专业怎么样？

3. I am sure to treasure the good chance and to study hard to realize my dream.
我一定会珍惜机会刻苦学习，实现我的梦想。

4. I am glad to hear that.
我很高兴听到你这么说。

5. By the way, what about the teaching facilities?

顺便问一下，你们学校的教学设施怎么样？

6. For vocational colleges, it is effective and practical to integrate theory with practice and to do practice geared to the needs of the jobs.

在高职院校，理论联系实际和对口实习是有效和实用的。

7. Would you please show me around your campus?

你愿意带我参观你们的校园吗？

Exercise 2: Complete the Following Dialogue in English

(A: an America visiting scholar   B: a Chinese student majoring in engineering cost)

A: How do you do?

B:_____?

（您好。）

A: May I ask you some questions?

B:_____.

（当然可以。）

A：What do you do?

B: _____.

（我是这个学校的一名学生。）

A: Can you tell me what major you are studying?

B: _____.

（工程造价专业。）

A: How do you think of this major?

B: _____.

（前景广阔，我喜欢。）

A: How many courses are there in this major?

B: _____.

（十多门课程，主要是专业必修课和基础课。）

A: Which course do you like best? And why ?

B: _____.

（建筑施工技术，因为老师讲课由浅入深，由易入难，并能理论联系实际。）

A: Thank you very much.

B: _____.

（不客气。）

Part Three: Reading Comprehension

Questions 1 to 5 are based on the following passage.

For years, high school students have received identical textbooks as their classmates. Even as students have different learning styles and abilities, they are force-fed the same materials. "Imagine a digital textbook where because I'm a different person and learn differently, my book is different from your book," said Richard Baraniuk, founder of OpenStax.

OpenStax will spend two years developing the personalized books and then test them on Houston-area students. The books will also go through a review and evaluation process similar to traditional textbooks. Baraniuk expects 60 people to review each book before publication to ensure its quality.

The idea is to make learning easier, so students can go on to more successful careers and lives. Baraniuk isn't just reproducing physical textbooks on digital devices, a mistake e-book publishers have made. He's seriously rethinking that the educational experience should be in a world of digital tools. To do this means involving individuals with skills traditionally left out of the textbook business. Baraniuk is currently hiring cognitive scientists and machine learning experts. Baraniuk wants to use the tactics（策略）of Google, Netflix and Amazon to deliver a personalized experience. These Web services all rely on complex algorithms（算法）to automatically adjust their offerings for customers.

Just as Netflix recommends different movies based on your preferences and viewing history, a textbook might present materials at a different pace. The textbook—which will be stored on a range of digital devices—will automatically adjust itself thanks to machine learning. As a student learns about a topic, he or she could be interrupted by brief quizzes that evaluate whether he or she masters the area. Depending on how the student does, the subject could be reinforced with more material. Or a teacher could be automatically emailed that the student is struggling with a certain concept and could use some one-on-one attention.

This personalized learning experience is possible thanks to the wealth of data a digital textbook can track. This data can be used to better track students' progress during a course. Parents and teachers can monitor a student's development and provide in time more proper assistance. With personalized learning methods, our students' talents will be better developed.

1. What do we learn about personalized books?

   A) Their quality will be ensured since they are developed by OpenStax.

B) They will be examined and judged before being published.

C) They will overlook different learning styles and abilities.

D) They will be much similar to traditional textbooks.

2. In which aspect have e-book publishers done incorrectly?

A) They have only put emphasis on learning experience.

B) They have made it difficult to have access to e-book.

C) They have made it rather boring and inconvenient to learn.

D) They have just produced an electronic copy of print textbooks.

3. What does Richard Baraniuk mean by "the educational experience should be in a world of digital tools" (Line 3-4, Para.3)?

A) Education should employ the machine to improve learning.

B) Education should involve traditional textbooks in the digital world.

C) Education should include obtaining skills by the use of machine learning.

D) Education should reproduce traditional textbooks on the Web services.

4. Personalized textbook is beneficial to the students because _____.

A) it stores the fixed material on different digital machines

B) it quizzes the student to make them more confident

C) it automatically presents movies based on the students' preference

D) it automatically matches learning material to the students' needs

5. Personalized learning experience may become possible owing to _____.

A) a great many digital equipment

B) the students' continuous progress

C) a great amount of digital information

D) parents' and teachers' constant watch

Part Four: Translation Skills

翻译技巧：增译法

常用的翻译技巧有增译法、省译法、重复法、转换法、拆句和合并法、正译法、反译法、倒置法、包孕法等。本章着重介绍增译法。

增译法指根据英汉两种语言不同的思维方式、语言习惯和表达方式，在翻译时增添一些单词、短语或句子，以便更准确地表达出原文所包含的意思。这种方式多半用在汉译英里。汉语中无主句的情况较多，而英语句子中则一般都要有主语，所以在翻译汉语无主句的时候，除了少数情况可用英语无主句、被动语态或"There be…"结构来翻译之外，一般都要根据语境补出主语，使句子更完整。

英汉两种语言在名词、代词、连词、介词和冠词的使用方法上也存在很大差别。英语中代词使用频率较高，凡说到人的器官和归某人所有的或与某人有关的事物时，必须在前面加上物主代词。因此，在汉译英时需要增补物主代词，而在英译汉时又需要根据情况适当删减。英语中词与词、词组与词组以及句子与句子的逻辑关系一般用连词来表示，而汉语则往往通过上下文和语序来表示这种关系。因此，在汉译英时常常需要增补连词。此外，英语句子也离不开介词和冠词。

在汉译英时还要注意增补一些原文中暗含而没有明言的词语以及一些概括性、注释性的词语，以确保译文意思的完整。总之，通过增译一是可以保证译文语法结构的完整，二是可以保证译文意思的明确。

例1： In the evening, after the banquet, the concert and table tennis exhibition, he would work on the drafting of the final communiqué.

翻译：晚上在参加宴会、出席音乐会、观看乒乓球表演之后，他还得起草最后公报。（增译动词）

详解：根据译文意思的需要，可以在名词前增加动词。比如把例1中的 after the banquet, the concert and table tennis exhibition 译为"在宴会、音乐会、乒乓球表演之后"，意思就不够明确，而如果在名词之前增加原文中虽无其词却有其意的动词，译为"在参加宴会、出席音乐会、观看乒乓球表演之后"，形成三个动宾词组，意思就明确了，读起来也较通顺自然，符合汉语习惯。

例2： Oh, Tom Canty, born in rags and dirt and misery, what sight is this!

翻译：哦，汤姆·康第，生在破烂、肮脏和苦难中，现在这番景象却是多么煊赫啊！（增译形容词）

详解：根据原著，汤姆·康第本是个贫儿，穿上王子服装以后，被人认为真的是王子，就显得特别煊赫。原文虽无"煊赫"的字眼，但含有此意，所以翻译时应增加上去。为了满足意思或修辞上的需要，对于一些英语句子中的名词，在翻译成汉语时可以增加一些适当的形容词，从而使语句更为通顺。

例3： In April, there was the "ping" heard around the world. In July, the ping "ponged".

翻译：四月里，全世界听到中国"乒"的一声把球打了出去；到了七月，美国"乓"的一声把球打了回来。（增译背景词语）

详解：翻译有时需要根据上下文及背景情况增加词语，比如例3如果直译就是"四月里全世界听到"乒"的一声；七月里，这"乓"声却"乓"了一下"，读者就会不知所云。同样的增译还可以用在中文成语和谚语的翻译上。例如"三个臭皮匠顶个诸葛亮。"增译注释性词语，为"Three cobblers with their wits combined equal Zhuge Liang the mastermind."

Translation Exercise 1

在大学，整个工科课程体系都非常强调数学的重要性。现在，数学包括统计学，其主要进行数据或者信息资料的收集、分类和使用工作。统计学的一个重要部分就是概率论，它研究的是当影响某个事件结果的不同因素或者变量存在时，某种结果出现的可能性。例如，在建造一座桥梁之前，要对该桥预承受的交通流量进行统计研究。在设计这座桥梁时，必须考虑诸如基础的水压力、冲击力、不同风力的作用以及其他很多因素。

Translation Exercise 2

建筑生产中有四大工种。这四大工种是土建工人、建筑安装工人、建筑机械工人和建筑装饰工人。当然，每一大工种都包括八种以上的工种。例如，建筑安装工种可分为水暖工、电工、焊工、起重工、通风工、铆工、安装钳工等。各工种之间密切配合，缺一不可。每一个工种都在建筑施工中起到重要的作用。

Unit 2 Design Process

Learning Objectives

After completing this unit, you will be able to do the following:
- √ Grasp the main idea and the structure of the text;
- √ Master the key language points and grammatical structure in the text;
- √ Understand the design process;
- √ Conduct a series of reading, listening, speaking and writing activities related to the theme of the unit.

Outline

The following are the main sections in this unit.
1. Warm-up Activity
2. Text
3. Words and Expressions
4. Exercises

Terms of Design Process

In this unit, you will learn the meanings of terms listed below.
diagram
budget
framework
phase

Vocabulary

Listed below are some words appearing in this unit that you should make them a part of your vocabulary.
blend
parallel
backtrack
novice
esoteric
intuitive
aesthetic
appreciation

Looking Ahead

The design process, also sometimes termed "problem solving process", includes a series of steps which usually (though not necessarily) follow a sequential order. In general terms, these same steps are also used by architects, industrial designers, engineers, and scientists to solve problems.

Introduction

The steps of the design process represent the ideal sequence of events. In reality, many of the steps overlap one another and blend together so the neat ordering of the outline is less clear or apparent. Further, some of the steps may parallel one another and occur simultaneously.

Engineers do not always follow the engineering design process steps one after another. It is very common to design something, test it, find a problem, and then go back to an earlier step to make a modification or change to the design. This way of working is called iteration, and it is likely that the process will do the same!

Warm-up Activity

What is needed to create a beautiful and practical design?

 ## Text

An Introduction to Design Process

Most good landscape architects employ a series of analytic and creative thinking steps referred to as the "design process" to arrive at a built site design that correctly meets all the necessary requirements in the most efficient and aesthetically pleasing manner. The design process has a number of uses including the following:

1. It provides a logical and organized framework for creating a design solution.

2. It helps to insure the solution that does evolve will be appropriately suited to the given circumstances of the design (the site, the client's needs, budget, etc.)

3. It aids in determining the best use of the land for the client by studying alternative solutions.

4. It serves as a basis of explaining and defending the design solution to the client.

The design process, also sometimes termed "problem solving process", includes a series of steps which usually (though not necessarily) follow a sequential order. In general terms, these same steps are also used by architects, industrial designers, engineers, and scientists to solve problems. For site designers, the design process typically includes these steps:

1. Client Contract

2. Research and Analysis (including site visit)

(1) Base Plan Preparation

(2) Site Inventory (Data Collection) and Analysis (Evaluation)

(3) Client Interview

(4) Program Development

3. Design

(1) Ideal Functional Diagram

(2) Site Related Functional Diagram

(3) Concept Plan

(4) Form Composition Study

(5) Preliminary Design

(6) Total Design

(7) Detailed Design

4. Implementation

(1) Construction

(2) Planting

5. Maintenance

6. Post-construction Evaluation

These steps of the design process represent the ideal sequence of events. In reality, many of the steps overlap one another and blend together so the neat ordering of the outline is less clear or apparent. Further, some of the steps may parallel one another and occur simultaneously. For example, Client Interview and Program Development may occur at the same time. One is also visiting the site and conducting a Site Analysis. At other times, one may find it necessary to backtrack to a precious step. As an illustration, one may find it necessary to revisit the site or talk to the client again once the design phase itself has been started because some items of information were overlooked the first time or one's memory and impression simply need refreshing.

It is important for a novice designer to understand the fact that beautiful and practical design solutions don't appear out of thin air like magic. There are no esoteric formulas or secret states of mind that produce good designs effortlessly. And designs are not created by only moving a pencil around on a piece of paper.

Designs that work well and affect our emotions require a great deal of sensitive observation, analysis, studying/thinking and restudying as well as some degree of inspiration and creativity.

It should be noted here that producing a design does involve both rational aspects (inventory, analysis, program development, construction knowledge) and intuitive aspects (the feel of putting forms and shapes together, aesthetic appreciation, etc.).The design process, then, is a framework of steps, incorporating both rational and intuitive phases, which aids the designer to organize his/her work, thoughts, and feelings in an effort to produce the best design solution possible.

Words and Expressions

landscape	['lændskeɪp]	n. 风景，景色，景观
analytic	[ˌænə'lɪtɪk]	adj. 分析的
aesthetically	[iːs'θetɪkəli]	adv. 美学观点上地
client	['klaɪənt]	n. 委托人，客户
framework	['freɪmwɜːk]	n. 框架，结构
evolve	[ɪ'vɒlv]	vt. 使发展，使进化，逐渐形成
budget	['bʌdʒɪt]	n. 预算
inventory	['ɪnvəntrɪ]	n. 财产目录，存货清单
sequential	[sɪ'kwenʃl]	adj. 连续的，有顺序的
diagram	['daɪəgræm]	n. 图表；简图
preliminary	[prɪ'lɪmɪnərɪ]	adj. 预备的，初步的
implementation	[ˌɪmplɪmen'teɪʃ(ə)n]	n. 实施
maintenance	['meɪntənəns]	n. 维护，保持
overlap	[ˌəʊvə'læp]	vt. 与……重叠
blend	[blend]	vt. 混合，把……混成一体
parallel	['pærəlel]	vt. 成平行，使……与……平行
backtrack	['bæktræk]	vt. 追踪，重做
novice	['nɒvɪs]	n. 新手
esoteric	[ˌesə'terɪk]	adj. 深奥的；秘传的
intuitive	[ɪn'tjuːɪtɪv]	adj. 直觉的，直观的
appreciation	[əˌpriːʃɪ'eɪʃ(ə)n]	n. 欣赏
incorporate	[ɪn'kɔːpəreɪt]	vt. 包含
phase	[feɪz]	n. 时期，阶段，方面
be referred to		被称为
be suited to		适合于
in general terms		概括地说

in reality 实际上
as an illustration 举例，例如
a great deal of 许多；大量

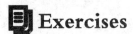 Exercises

Part One : Special Terms

1. qualitative analysis　　　　　_____
2. skeleton framework　　　　　_____
3. engrave a design on metal　_____
4. 惊人的相似之处　　　　　　_____
5. 定量分析　　　　　　　　　_____
6. 金属框架　　　　　　　　　_____

Part Two: Situational Conversation

Jordan: Hi, can you do me a favor, Engineering Li?

Li Can: At your service.

Jordan: I don't know the exact position of the switch board. Will you scratch it on the wall?

Li Can: It should be scratched in accordance with your drawing. Where is the drawing?

Jordan: Here it is, Engineering Li.

Li Can: Sorry. You have got a wrong drawing.

Jordan: What is the problem?

Li Can: It is a plan, but what I need is an elevation, where the switch board is clearly marked. Would you please get it for me now?

Jordan: OK. Here you are.

Li Can: Thank you. (After a while) Look, the switch board is near the corner.

Jordan: You are very observant. I do admire you very much.

Li Can: If you were in my position, you'd have done the same, I am sure.

Jordan: That's true. But how high is the switch board to the floor?

Li Can: It is 1.5 m.

Jordan: How do you know the exact size?

Li Can: Because what you gave me for the second time is a dimensioned drawing. If you don't believe it, you can get a drawing scale to measure it yourself.

Jordan: In fact, there is no need at all to do that. So reading and studying carefully is an essential prerequisite for construction well, isn't it?

Li Can: Yes, absolutely. As every drawing is meticulously designed, the size and figures on the drawing are also calculated exactly, we constructors should regard it as the norm and the guide of our construction works.

Jordan: That means construction is the translation of design into reality, am I right?

Li Can: Quite right. It seems that you don't read the drawing well.

Jordan: Oh, just so so.

Li Can: Personally, I think you should make effort to read the drawing well. Only in this way, can you do your work well. Don't you think so?

Jordan: Sure. I will follow your advice and redouble my efforts to make up for knowledge of building drawings.

Li Can: If you have any drawing problem in your future work and study, please don't hesitate to call me, and I will do my best to help you.

Jordan: You are very kind. Thank you very much.

Li Can: Not at all. I am honored to do something useful for you.

Exercise 1: Sentence Patterns

1. Will you do me a favor to lift the heavy box?
你能帮我提这个重箱子吗？

2. What is the problem/What is wrong/What is the matter with the machine?
机器出了什么问题？

3. If you were in my position, you'd have done the same, I am sure.
我相信如果你处在我的位置上，你也会这么做。

4. Personally, I think you should make effort to read the drawing well.
我个人觉得你应该努力学习识图。

5. Only in this way, can you do your work well.
只有这样，你才能做好你的工作。

6. I will follow your advice and redouble my efforts to make up for knowledge of building drawings.
我会听取您的建议，加倍努力补习关于建筑图纸方面的知识。

7. If you have any questions, don't hesitate to ask.

有什么问题只管问。

8. I am honored to do something useful for you.

能为你做些有用的事情我深感荣幸。

Exercise 2: Complete the Following Dialogue in English

A: Excuse me! Can you answer me some questions about building drawings?

B:_____.

（当然可以。关于建筑图纸你想了解什么？）

A: First, I want to know the scientific definition of building drawings.

B:_____.

（所谓的建筑图纸就是将一幢拟建建筑物的内外形状和大小以及各部分的结构、构造、装饰、设备等内容，按照有关规范规定，用正投影方法详细准确地画出的图样。）

A: So building drawings are very important to the construction works, aren't they?

B:_____.

（你说的没错。建筑图纸一直被看作施工的指南和准则。）

A: How many parts are there in a complete set of building drawings?

B:_____.

（一套完整的施工图，根据专业内容或者作用不同，一般包括图纸目录、设计总说明、建筑施工图、结构施工图以及设备施工图。）

A: How can I tell the difference between the building construction drawings and the structural working drawings?

B:_____.

（建筑施工图表示的是建筑物的内部布置情况、外部形状以及装修、构造、施工要求等。）

A: What about structural working drawings?

B:_____.

（结构施工图表示的是承重结构的布置情况、构件类型、尺寸、大小及结构做法等内容。）

A: To make these complicated drawings, some precise instruments are inevitable, am I right?

B:_____.

（是的，常用的绘图仪器有绘图笔、绘图圆规、丁字尺、曲线板等。）

A: I know. They are mainly used in manual drawings, aren't they?

B:_____.

（没错，现在的建筑图纸主要是通过 AutoCAD 软件绘制出来的。相对于手工绘图，其优点是不言而喻的。）

A: What is AutoCAD, I want to know?
B:_____.

（AutoCAD 是美国 Autodesk 公司开发的用于二维及三维设计、绘图的自动计算机辅助软件。）

A: Would you mind teaching me to make some drawings by AutoCAD when you are free?
B:_____.

（当然不介意，如有需要，尽管给我打电话，我一定会不遗余力地帮你。）

Part Three: Reading Comprehension

Questions 1 to 5 are based on the following passage.

A new study shows that students learn much better through an active, iterative（反复的）process that involves working through their misconceptions with fellow students and getting immediate feedback from the instructor.

The research was conducted by a team at the University of British Columbia (UBC), Vancouver, in Canada, led by physics Nobelist Carl Wieman. In this study, Wieman trained a postdoc, Louis Deslauriers, and a graduate student, Ellen Schelew, in an educational approach, called "deliberate practice", that asks students to think like scientists and puzzle out problems during class. For one week, Deslauriers and Schelew took over one section of an introductory physics course for engineering majors, which met three times for one hour. A tenured physics professor continued to teach another large section using the standard lecture format. The results were dramatic: After the intervention, the students in the deliberate practice section did more than twice as well on a 12-question multiple-choice text of the material as did those in the control section. They were also more engaged and a post-study survey found that nearly all said they would have liked the entire 15-week course to have been taught in the more interactive manner.

"It's almost certainly the case that lectures have been ineffective for centuries. But now we've figured out a better way to teach that makes students an active participant in the process," Wieman says. The "deliberate practice" method begins with the instructor giving students a multiple-choice question on a particular concept, which the students discuss in small groups before answering electronically. Their answers reveal their grasp of the topic, which the instructor deals with in a short class discussion before repeating the process with the next concept.

While previous studies have shown that this student-centered method can be more effective than teacher-led instruction, Wieman says this study attempted to provide "a particularly clean comparison...to measure exactly what can be learned inside the classroom." He hopes the study persuades faculty members to stop delivering traditional lecture and "switch over" to a more interactive approach. More than 55 courses at Colorado across several departments now offer that approach, he says, and the same thing is happening gradually at UBC.

1. What can we know about the study led by Carl Wieman in the second paragraph?

    A) Students need to turn to scientists for help if they have trouble.

    B) An introductory physics course was given to physics majors.

    C) Students were first taught by the "deliberate practice" approach.

    D) A professor continued to teach the same section with the traditional lectures.

2. The results of the research reveal that _____.

    A) the students in the experimental section performed better on a test

    B) the students in the control section seemed to be more engaged

    C) the students preferred the traditional lectures to deliberate practice

    D) the entre 15-week course was actually given in the new manner

3. How does Wieman look at the traditional lectures according to the third paragraph?

    A) They have lasted for only a short period of time.

    B) They continue to play an essential role in teaching.

    C) They can make students more active in study.

    D) They have proved to be ineffective and outdated.

4. How does the "deliberate practice" method work?

    A) The students are first presented with some open questions.

    B) The students have to hand in paper-based homework.

    C) The instructor remains consistent in the way of explaining concept.

    D) The instructor expects the students to air their views at any time.

5. We learn from the last paragraph that Wieman's new approach _____.

    A) can achieve the same effects as the traditional lecture format

    B) can evaluate the student's class performance roughly

    C) will take the place of the traditional way of teaching in time

    D) has been accepted by faculty members in some colleges

Part Four: Translation Skills

翻译技巧：省译法

省译法是指在翻译中舍去原文中那些可有可无的成分。省译目的在于使译文更加流畅，符合目的语的习惯。英译汉时省略的大多是英语中因为语法上的需要而存在，但根据汉语习惯并不需要译出来的词语。因此，省译的原则是：省译的部分在译文中不言而喻，译出来反而拖沓累赘；省译后不能有损原意或改变原文色彩。

具体而言，省译法包括省略原文中的词语、省略语法范畴以及简化累赘的文体。这是与增译法相对应的一种翻译方法，即删去不符合目标语思维习惯、语言习惯和表达方式的词，以避免译文累赘。如：

例1： He puts his hand into his pockets and then shrugged his shoulders.

翻译：他把双手放进口袋里，然后耸耸肩。（省译代词）

详解：本例句如果直译出来是"他把他的双手放进了他的口袋里，然后耸耸他的肩"。这样翻译就显得语句有些拖沓、累赘，因此可以省译三个物主代词"his"，译为"他把双手放进口袋里，然后耸耸肩。"这样意思明确，读起来也较通顺、自然，符合汉语习惯。

例2： The report began, and the audience stopped talking.

翻译：报告开始了，听众停止了谈话。（省译连词）

详解：英语中语句的逻辑关系往往是通过连词明显地表现出来的，然而中文语句间的关系往往是内隐的。因此英译汉时，就需要适当省译连词。如例2中的"and"，这种顺接关系在汉语中是不需要明确表现出来的，故省译该连词。

例3： He hit her in the face.

翻译：他打了她的脸。（省译介词）

详解：有时根据汉语习惯并不需要译出介词，比如例3如果直译就是"他打进了她的脸"，这样读者就会不知所云。所以此处需要省译介词"in"，译作"他打了她的脸"。

Translation Exercise 1

建筑图纸有许多种，如平面图、立面图、剖面图、平面布置图、鸟瞰图、仰视图、标准图、施工图……图纸在建筑施工中起着至关重要的作用。因此，图纸被视为施工的准则和指南，也就是说，施工就是把设计转为现实。

Translation Exercise 2

对于承包人，为竞标或谈判而递交给业主的投标报价包括现场管理费用在内的直接施工成本，加上包含常规企业管理费用和利润在内的上涨幅度。投标报价中的直接施工成本通常根据以下因素而得出：分包商报价、工程量清单和施工程序。

Unit 3   Structural Materials

Learning Objectives

After completing this unit, you will be able to do the following:
- √ Grasp the main idea and the structure of the text;
- √ Master the key language points and grammatical structure in the text;
- √ Understand the basic knowledge of structural materials;
- √ Conduct a series of reading, listening, speaking and writing activities related to the theme of the unit.

Outline

The following are the main sections in this unit.
1. Warm-up Activity
2. Text
3. Words and Expressions
4. Exercises

Terms of Structural Materials

In this unit, you will learn the meanings of terms listed below.
- structural materials
- plain (unreinforced) concrete
- prestressed concrete
- reinforced concrete
- plywood
- laminated wood
- Portland cement
- glass-fiber-reinforced composite

Vocabulary

Listed below are some words appearing in this unit that you should make them a part of your vocabulary.
- availability
- concrete
- smelt
- malleable
- cement
- patent
- vigorous
- heterogeneous
- tensile
- compressive
- particle

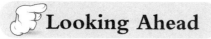

Structural Materials

Structural materials encompass materials whose primary purpose is to transmit or support a force. Applications can be in transportation (aircraft and automobiles), construction (buildings and roads), or in components used for body protection (helmets and body armor), energy production (turbine blades), or other smaller structures such as those used in microelectronics. Structural materials can be metallic, ceramic, polymeric or a composite between these materials. This unit will focus on the structural materials for construction.

Introduction

Structural engineering depends on the knowledge of materials and their properties, in order to understand how different materials support and resist loads.

Common structural materials are as follows.

1. Iron: Wrought iron, cast iron, steel, stainless steel;

2. Concrete: reinforced concrete, prestressed concrete;

3. Aluminum;

4. Composites;

5. Masonry;

6. Timber; and other structural materials are adobe, alloy, bamboo, carbon fibre, fiber reinforced plastic, mudbrick, roofing materials, and so on.

Warm-up Activity

Can you point out the structural materials of each picture? Which one is reinforced and prestressed concrete? And which one is sandwich panels?

Can you image what the structural materials would be in the future?

Text

The availability of suitable structural materials is one of the principal limitations on the accomplishment of an experienced structural engineer. Early builders depended almost exclusively on wood, stone, brick, and concrete. Although iron had been used by humans at least since the building of the Egyptian pyramids, use of it as a structural material was limited because of the difficulties of smelting it in large quantities. With the industrial revolution, however, came both the need for iron as a structural material and the capability of smelting it in quantity.

John Smeaton, an English civil engineer, was the first to use cast iron extensively as a structural material in the mid-eighteenth century. After 1841, malleable iron was developed as a more reliable material and was widely used. Whereas malleable iron was superior to cast iron, there were still too many structural failures and there was a need for a more reliable material. Steel was the answer to this demand. The invention of the Bessemer converter in 1856 and the subsequent development of the Siemens-Martin open-hearth process for making steel made it possible to produce structural steel at competitive prices and triggered the tremendous developments and accomplishments in the use of structural steel over the next hundred years.

The most serious disadvantage of steel is that it oxidizes easily and must be protected by paint or some other suitable coating. When steel is used in an enclosure where a fire could occur, the steel members must be encased in a suitable fire-resistant enclosure such as masonry, concrete. Normally, steel members will not fail in a brittle manner unless an unfortunate combination of metallurgical composition, low temperature, and bi-or triaxial stress exists.

Reinforced and prestressed concrete share with structural material. Natural cement concrete have been used for centuries. Modern concrete construction dates from the middle of the nineteenth century, though artificial Portland cement was patented by Aspidin, an Englishman, about 1825. Modern cement is a mixture of limestone and clay, which is heated and then ground into a powder. It is mixed at or near the construction site with sand, aggregate (small stones, crushed rock, or gravel), and water. Different proportions of the ingredients produce concrete with different strength and weight. Although several builders and engineers experimented with the use of steel-reinforced concrete in the last half of the nineteenth century, its dominant use as a building material dates from the early decades of the twentieth century. The last fifty years have seen the rapid and vigorous development of prestressed concrete design and construction, founded largely on early work by Freyssinet in France and Magnel in Belgium.

Plain (unreinforced) concrete not only is a heterogeneous material but also has one very serious defect as a structural material, namely, its very limited tensile strength, which is only of the order of one-tenth its compressive strength. Not only is tensile failure in concrete of a brittle type, but likewise compression failure occurs in a relatively brittle fashion without being preceded by the forewarning of large deformations. (Of course, in reinforced-concrete construction, ductile behavior can be obtained by proper selection and arrangement of the reinforcement.) Unless proper care is used in the selection of aggregates and in the mixing and placing of concrete, frost action can cause serious damage to concrete masonry. Concrete creeps under long-term loading to a degree that must be considered carefully in selecting the design stress conditions. During the curing process and its early life, concrete shrink a significant amount, which to a degree can be controlled by properly proportioning the mix and utilizing suitable construction techniques.

With all these potentially serious disadvantages, engineers have learned to design and build beautiful, durable, and economical reinforced concrete structures for practically all kinds of structural requirements. This has been accomplished by careful selection of the design dimensions and the arrangement of the steel reinforcement, development of proper cements, selection of proper aggregates and mix proportions, careful control of mixing, placing, and curing techniques and imaginative development of construction methods, equipment and procedures.

The versatility of concrete, the wide availability of its component materials, the unique ease of shaping its form to meet strength and functional requirements, together with the exciting potential of further improvements and development of not only the newer prestressed and precast concrete construction but also the conventional reinforced concrete construction, combine to make concrete a strong competitor of other materials in a very large fraction of structures. Prestressed concrete is an improved form of reinforcement. Steel rods are bent into the shapes to give them the necessary degree of tensile strength. They are then used to prestress concrete, usually by pretensioning or posttensioning method. Prestressed concrete has made it possible to develop buildings with unusual shapes, like some of the modern sports arenas, with large spaces unbroken by any obstructing supports.

In modern times, with the increased use of steel and reinforced concrete construction, wood has been relegated largely to accessory use during construction, to use in temporary and secondary structures, and to use for secondary members of permanent construction. Modern technology in the last sixty years has revitalized wood as a structural material, however, by developing vastly improved timber connectors, various treatments to increase the durability of wood, and laminated wood made of thin layers bonded together with synthetic glues using revolutionary gluing techniques. Plywood with essentially nondirectional strength properties is the most widely used laminated wood, but techniques have also been developed for building large laminated wood members that for certain structures are competitive with concrete and steel.

Materials with future possibilities are the engineering plastics and the exotic metals and their alloys, such as beryllium, tungsten, tantalum, titanium, molybdenum, chromium, vanadium, and niobium. There are many different plastics available, and the mechanical properties exhibited by this group of materials vary over a wide range that encompasses the range of properties available among the more commonly used structural materials. Thus in many specific design applications it is possible to select a suitable plastic material for an alternative design. Experience with the use of plastics outdoors is limited. Generally speaking, however, plastics must be protected from the weather. This aspect of design is therefore a major consideration in the use of plastics for primary structural elements. One of the most promising potential used of plastics is for panel and shell-type structures.

Laminated or sandwich panels have been used in such structures with encouraging results that indicate an increased use in this type of construction in the future.

Another materials development with interesting possibilities is that of composites consisting of a matrix reinforced by fibers or fiber-like particles. Although glass-fiber-reinforced composites with a glass or plastic matrix have been used for years, they appear to have much broader possibilities for a large variety of secondary structural components. Fiber-reinforced concrete is another composite being actively studied and developed. Several experimental applications are being observed under service conditions. Experiments have been conducted with both steel and glass fibers, but most of the service experience has been with steel fibers.

Words and Expressions

structural	['strʌktʃ(ə)r(ə)l]	*adj.* 结构的；构造的；建筑的；建筑用的
availability	[əˌveɪlə'bɪləti]	*n.* 有效性；可用性；可得到的人（或物）；可用性
concrete	['kɒŋkriːt]	*adj.* 具体的；实质性的；混凝土的 *n.* 水泥；混凝土 *v.* 凝结
smelt	[smelt]	*vt.* 熔炼
malleable	['mælɪəb(ə)l]	*adj.* 可塑的；易适应的；有延展性
trigger	['trɪgə]	*v.* 使发生；触发；使运行 *n.* 扳机；起因
oxidize	['ɒksɪdaɪz]	*v.* 氧化；生锈
masonry	['meɪs(ə)nri]	*n.* 石造建筑；石工行业
brittle	['brɪt(ə)l]	*adj.* 易碎的；尖利的；脆弱的
metallurgical	[ˌmetə'lɜːdʒɪkl]	*adj.* 冶金的；冶金学的
reinforced	[riːɪn'fɔːst]	*adj.* 加强的
prestress	[prɪ'stres]	*vt.* 对……预加应力 *n.* 预先拉伸
posttensioning	[pəʊst'tenʃənɪŋ]	*n.* 后张（后加拉力的）
cement	[sɪ'ment]	*n.* 水泥；纽带；接合剂；牙骨质；补牙物；基石；*vt.* 用水泥涂；巩固 *vi.* 接合起来
patent	['pæt(ə)nt]	*n.* 专利；特许 *vt.* 授予专利 *adj.* 专利的

vigorous	['vɪg(ə)rəs]	*adj.* 精力充沛的；有力的；元气旺盛的
heterogeneous	[ˌhet(ə)rə(ʊ)'dʒiːnɪəs]	*adj.* 异种的；异质的；由不同成分形成的
tensile	['tensaɪl]	*adj.* 可拉长的；可伸展的；张力的
compressive	[kəm'presɪv]	*adj.* 有压缩力的；压缩的
particle	['pɑːtɪk(ə)l]	*n.* 颗粒；微粒；极小量；质点；粒子
matrix	['meɪtrɪks]	*n.* 矩阵；发源地；基质；母体；子宫；（人或社会成长发展的）社会环境；政治局势线路网；道路网；[地]杂基；矩阵转接电路；唱片模板
cast iron		铸铁

Exercises

Part One : Special Terms

1. tensile strength　　　　　　　　　　_____
2. compressive strength　　　　　　　　_____
3. reinforced and prestressed concrete　_____
4. 素混凝土　　　　　　　　　　　　　_____
5. 预应力混凝土　　　　　　　　　　　_____
6. 波特兰水泥（硅酸盐水泥）　　　　　_____

Part Two : Situational Conversation

(**Mark:** a wholesaler　**Li Lu:** an old customer)

Mark: We haven't seen each other for a long time, Mrs. Li. What can I do for you?

Li Lu: Long time no see! I am here to buy some building materials.

Mark: What materials do you want to buy?

Li Lu: Mainly common materials, such as cement, steel bar and timber. Can you tell me their current market prices?

Mark: Oh, let me have a look. Cement is 1 dollar per kilogram, steel bar is 700 dollars per ton, and timber 650 dollars per cubic meter.

Li Lu: Could you show me the price catalogue?

Mark: Ok, here it is. The materials here are not expensive, but of the best quality and can be guaranteed. You can rest assured.

Li Lu: Honestly speaking, the prices are much higher than they were a few years ago. Do you have any specials for frequent buyers like me?

Mark: To tell you the truth, the prices of the building materials keep going up recently, which make it difficult to return good profits for frequent buyers like you. We can offer varying degree of discount depending on the amount of your purchase.

Li Lu: I want to know the most favorable price you can offer for 20 tons of cement, 1,000 tons of steel bar and 1,000 cubic meters of timber.

Mark: In consideration of the quantity of the three building materials you want to purchase, we can offer you 10% discount.

Li Lu: 10%? It's too low a rate. Could you see your way to increase it to 30%?

Mark: You want to drive me bankrupt? You can't expect us to make a large reduction.

Li Lu: Oh, with such a large order on hand, you needn't to worry anymore. Please think it over, my old friend.

Mark: Considering the long-standing business relationship between us, we shall grant you a special discount of 10%. As you know, we do business on the basis of equality and mutual benefit. Besides, the price of the building materials tends to go up and there is a heavy market demand of them.

Li Lu: Yes, I know the present market situations. Anyhow, let's meet each other halfway, how about 15%?

Mark: You are a real businesswoman! All right, I agree to give you a 15% discount.

Li Lu: Deal! But keep your words and deliver the materials in time!

Mark: Make yourself easy. We will go all-out to supply you with the materials and deliver them to your site without any delays.

Li Lu: Thank you!

Notes:

1. common materials 普通材料，大众材料。

2. current market price 当前市场价格。

3. honestly speaking 坦诚地说，类似的还有 frankly speaking 坦白地说, generally speaking 总体说来。

4. To tell you the truth 为插入语，意思为"老实说，说实话"。

5. which make it difficult to return good profits 是非限定性定语从句，意思是"使得我们很难有好的收益"。

6. cubic meters 立方米，ton 吨。

7. cement 水泥，steel bar 钢筋，timber 木材。

8. go all-out 全力以赴，鼓足干劲。

Exercise 1: Sentence Patterns

1. We haven't seen each other for a long time. 我们很久没有见面了。

2. Mainly common materials, such as cement, steel bar and timber. 主要是一些大众材料，如水泥、钢筋还有木材。

3. Can you tell me their current market prices? 能告诉我它们的当前市场价格吗？

4. Cement is 1 dollar per kilogram, steel bar is 700 dollars per ton, and timber 650 dollars per cubic meter. 水泥每千克1美元，钢筋每吨700美元，木材是每立方米650美元。

5. Could you show me the price catalogue? 可以把价格目录表给我看一下吗？

6. The materials here are not expensive, but of the best quality which can be guaranteed. You can rest assured. 这里的材料不贵、质量最好、有保证。您可以放心。

7. We can offer varying degree of discount depending on the amount of your purchase. 根据您购买的数量我们可以给予不同程度的折扣。

8. Honestly speaking, the prices are much higher than they were a few years ago. 说实话，这价格比几年前高了不少。

9. Keep your words and deliver the materials in time! 请兑现您的承诺，按时送达材料！

Exercise 2: Complete the Following Dialogue in English.

(A: a salesman   B: a customer)

A: Can I help you?

B:_____.

（嗯，我打算订购一批石材。）

A: For what purpose?

B:_____.

（家里装修用的。）

A: All the samples of stone materials are here. What do you want to buy?

B:_____.

（我对你这里的大理石很满意，如果价格合理的话，我现在就订货。）

A: I'm very glad to hear that.

B:_____?

（这种纯白的大理石最低价是多少？）

A: 700 yuan per square meter.

B: I think the price is too high. Can you reduce it?

A:_____.

（这恐怕不行，700元是我们的底价了。更何况这种大理石是纯天然的，每一块都有独一无二的图案和色彩。）

B: But there are some beautiful marbles in the market priced at 120-180 yuan per square meter. How do you explain the large price differences to me?

A:_____.

（那些价格低的是人造大理石，透明度不好，且没有光泽。）

B: Oh, I got it. Can you tell me how to distinguish the two different kinds of marbles?

A:_____.

（最简单的方法就是滴上几滴稀盐酸，天然大理石会剧烈起泡，人造大理石起泡弱甚至不起泡。）

B: Can I do a small experiment on the samples of marbles here by using diluted hydrochloric acid?

A:_____.

（当然可以。真金不怕火炼。）

(After the experiment)

B: Well, I'll accept the price and place an initial order of 10,000 square meters.

A:_____.

（太好了。跟你做生意真是我的荣幸。）

B: The pleasure is ours. Can you deliver the goods by June 2nd?

A:_____.

（绝对没有问题，我们不会误事的。）

Part Three: Reading Comprehension

That Orientals and Westerners think in different ways is not mere prejudice. Many psychological students conducted over the past two decades suggest Westerners have a more individualistic and abstract mental life than East Asians do. Several expectations are proposed to account for different ways of thinking.

One explanation is that stepping into the modern social, economic and technological situation promotes individualism. However, in Japan, a pretty is liable to a higher frequency of infectious disease, it is more dangerous to make contact with strangers, which causes groups in this place to turn inward and tend to be collective. This explanation is also been questioned. Europe has had its share of plagues; probably more that either Japan or Korea. And though in southern China, a source of infection often stars there, this is not true of other parts of that enormous country.

That led Thomas of the university of Virginia and his colleagues to look into a third suggest: that the crucial difference is agricultural. The western staple is wheat, while the Eastern is rice. Before that the crucial difference is agricultural, a farmer who grew rice had to spend twice as many hours doing so as one who grew wheat. To promote efficient agricultural production, especially at times of painting and harvesting, rice-growing societies as far apart as India, Malaysia and Japan all developed cooperative labor exchanges. That is, neighbors arranged their farms schedules one after another in order to assist each other during these significant periods. Since, until recently, almost everyone was a farmer, it is a reasonable proposal that such a collective outlook would enjoy a controlling position in society culture and behavior, and might prove so deep-rooted that even now, when most people earn their living in other ways, it helps to define their lives. This proposal that the different ways of thinking of East and West are, at least in part, a consequence of their agriculture is worth future exploration.

1. What do we learn about Easterners way of thinking?

    A) They often hold no prejudice.   B) They pursue abstract mental life.

    C) They are in favor of individualism.   D) They emphasize the collective values.

2. What does the first explanation indicate?

    A) Individualism has little relationship with the modern development.

    B) Individualism is inspired by modern society, economy and technology.

    C) Japan can be counted as a modern country.

    D) The Japanese prefer a more collective life.

3. Which of the following questions is the second proposal?

    A) People in a place with infectious disease tend to advocate individualism.

    B) People in a place with infectious disease tend to contact with outsiders.

    C) In China, it is easy to prevent the spread of infectious disease.

    D) Europe is more likely to be infected by plagues than Korea.

4. What do Thomas and his colleagues assume?

    A) The cultivation of wheat requires twice as much time as that of rice.

    B) In Malaysia, farmers need their neighbors assistance to grow wheat.

    C) The major agricultural crop of rice may define cooperative outlook.

    D) Farmers are only engaged in their own planting and harvesting of rice.

5. The author attitude toward Thomas and his colleagues suggestion is _____.
   A) indifferent   B) objective   C) critical   D) disappointed

Part Four: Translation Skills
翻译技巧：重复法

重复法（repetition）是在翻译中为了使译文忠实于原文并且产生意义明确、文字通顺、流畅，符合目的语习惯的文字，而将某一部分文字反复使用的翻译技巧。在译文中适当地重复原文中出现过的词语，以使意思表达得更加清楚；或者进一步加强语气，突出强调某些内容，都会收到更好的修辞效果。一般情况下，英语常常会为了行文简洁而尽量避免重复。因此，经常借助替代、省略或变换等其他表达方法。与此相反，重复是汉语表达的一个显著特点。在许多场合某些词语不仅需要重复，而且也只有重复这些词语，语义才能明确，表达才能生动。为了达到汉语译文准确、通顺和完整的翻译标准，在英译汉中，经常会采用重复法，如：

例1： You must ask the mother at home, the children in the street, the ordinary man in the market and look at their mouth, how they speak, and translate that way; then they'll understand and see that you're speaking to them in German.

翻译：你一定要问一问家庭主妇，问一问街头玩耍的孩子，问一问集市上做买卖的百姓，听听他们说些什么，他们如何说，你就如何译；这样他们就会理解，就会明白：你是在用德语和他们讲话。（重复谓语动词）

详解：该句的中文翻译将原文的谓语动词 ask 进行了三次重复，由于 ask 的宾语各不相同，如果不将谓语部分重复表达，会影响"问一问"宾语对象的明确性。

例2： We have to analyze and solve problems.

翻译：我们必须分析问题，解决问题。（重复宾语）

详解：在本句的翻译中，将句子中的宾语 "problems" 进行了重复翻译，其目在于能够将 analyze 和 solve 的宾语交代清楚，如果不加以重复，可能会对分析的对象产生疑问。

例3： Ignorance is the mother of fear as well as of admiration.

翻译：无知是畏惧之源，羡慕之根。（重复省略的部分）

详解：本句的英文原文中有一个省略成分，也就是省略了 as well as 后面的 mother，在英语中，这种省略方式可以避免表达上的重复和拖沓。而在中文中如果采取同样的方式，就会译为 "无知是畏惧也是羡慕之根"，会产生一定的歧义。因此，为避免不必要的歧义，应对省略部分进行重复翻译。

Translation Exercise 1

水泥与钢筋混凝土的联合委员会于1904年建立，包括美国土木工程师协会、美国测试和材料协会、美国铁路工程协会和美国波特兰水泥制造协会。该团体中后来又加入了美国混凝土协会。在1904至1910年，联合委员会进行了研究。在1913年发布的初步报告中列出了在1898至1911年发表的关于钢筋混凝土的重要文件和书籍。该委员会最后的报告发表于1916年。Kerekes和Reid在1954年对钢筋混凝土建筑规范在美国的发展历史作了回顾。

Translation Exercise 2

屋架组拼：屋架分片运至现场组装时，拼装平台应平整。组装时应保证屋架总长及起拱尺寸的要求。焊接时焊完一面检查合格后，再翻身焊另一面，并做好施工记录。经验收合格后方准吊装。屋架及天窗架也可以在地面上组装好一次吊装，但要临时加固，以保证吊装时有足够的刚度。

Unit 4　Bridges

Learning Objectives

After completing this unit, you will be able to do the following:
- √ Grasp the main idea and the structure of the text;
- √ Master the key language points and grammatical structure in the text;
- √ Understand the basic knowledge of bridges;
- √ Conduct a series of reading, listening, speaking and writing activities related to the theme of the unit.

Outline

The following are the main sections in this unit.
1. Warm-up Activity
2. Text
3. Words and Expressions
4. Exercises

Terms of Bridges

In this unit, you will learn the meanings of terms listed below.
beam bridge
truss bridge
cantilever bridge
arch bridge
tied arch bridge
suspension bridge
cable-stayed bridge

Vocabulary

Listed below are some words appearing in this unit that you should make them a part of your vocabulary.
terrain
crossbeam
span
anchor
log
plank
cavern
intermediate
timber
masonry
cantilever
suspension

Looking Ahead

Bridge

Planning and designing of bridges is part art and part compromise, the most significant aspect of structural engineering. It is the manifestation of the creative capability of designers and demonstrates their imagination, innovation, and exploration. Bridge design is a complex engineering problem. The design process includes consideration of other important factors, such as choice of bridge system, materials, dimensions, foundations, aesthetics, and local landscape and environment.

Introduction

Bridge is a structure built to span a valley, road, river, body of water, or any other physical obstacle. Designs of bridges will vary depending on the function of the bridge and the nature of the area where the bridge is to be constructed.

There are seven main types of bridges: beam bridges, cantilever bridges, arch bridges, tied arch bridges, suspension bridges, cable-stayed bridges and truss bridges.

Warm-up Activity

Do you know which bridge is Golden Gate Bridge and which one is Tower Bridge? Can you list other famous bridges? What kind of bridges do you want to design?

Text

A bridge is a structure built to span physical obstacles such as a body of water, valley, or road, for the purpose of providing passage over the obstacle. There are many different designs that all serve unique purposes and apply to different situations. Designs of bridges vary depending on the function of the bridge, the nature of the terrain where the bridge is constructed and anchored, the material used to make it, and the funds available to build it.

Original bridges made by humans were probably spans of cut wooden logs or planks and eventually stones, using a simple support and crossbeam arrangement. Some early Americans used trees or bamboo poles to cross small caverns or wells to get from one place to another. A common form of lashing sticks, logs, and deciduous branches together involved the use of long reeds or other harvested fibers woven together to form a connective rope with the capability of binding and holding together the materials used in early bridges.

Types of Bridges

Bridges can be categorized in several different ways. Common categories include the type of structural elements used, by what they carry, whether they are fixed or movable, and by the materials used.

Structure type

Bridges may be classified by how the forces of tension, compression, bending, torsion and shear are distributed through their structure. Most bridges will employ all of the principal forces to some degree, but only a few will predominate. The separation of forces may be quite clear. In a suspension or cable-stayed span, the elements in tension are distinct in shape and placement. In other cases the forces may be distributed among a large number of members, as in a truss, or not clearly discernible to a casual observer as in a box beam.

Beam bridge	Beam bridges are horizontal beams supported at each end by substructure units and can be either simply supported when the beams only connect across a single span, or continuous when the beams are connected across two or more spans. When there are multiple spans, the intermediate supports are known as piers. The earliest beam bridges were simple logs that sat across streams and similar simple structures. In modern times, beam bridges can range from small wooden beams to large steel boxes. The vertical force on the bridge becomes a shear and flexural load on the beam which is transferred down its length to the substructures on either side. They are typically made of steel, concrete or wood. Beam bridge spans rarely exceed 250 feet (76 m) long, as the flexural stresses increase proportional to the square of the length (and deflection increases proportional to the 4th power of the length). However, the main span of the Rio-Niteroi Bridge, a box girder bridge, is 300 metres (980 ft). The world's longest beam bridge is Lake Pontchartrain Causeway in southern Louisiana in the United States, at 23.83 miles (38.35 km), with individual spans of 56 feet (17 m). Beam bridges are the most common bridge type in use today.

 **Truss bridge**	A truss bridge is a bridge whose load-bearing superstructure is composed of a truss. This truss is a structure of connected elements forming triangular units. The connected elements (typically straight) may be stressed from tension, compression, or sometimes both in response to dynamic loads. Truss bridges are one of the oldest types of modern bridges. The basic types of truss bridges shown in this article have simple designs which could be easily analyzed by the nineteenth and the early twentieth century engineers. A truss bridge is economical to construct owing to its efficient use of materials.
 **Cantilever bridge**	Cantilever bridges are built using cantilevers-horizontal beams supported on only one end. Most cantilever bridges use a pair of continuous spans that extend from opposite sides of the supporting piers to meet at the center of the obstacle the bridge crosses. Cantilever bridges are constructed using much the same materials and techniques as beam bridges. The difference comes in the action of the forces through the bridge. Some cantilever bridges also have a smaller beam connecting the two cantilevers, for extra strength. The largest cantilever bridge is the 549-metre (1,801 ft) Quebec Bridge in Quebec, Canada.
 **Arch bridge**	Arch bridges have abutments at each end. The weight of the bridge is thrust into the abutments at either side. The earliest known arch bridges were built by the Greeks, and include the Arkadiko Bridge. With the span of 220 m (720 ft), the Solkan Bridge over the Soča River at Solkan in Slovenia is the second largest stone bridge in the world and the longest railroad stone bridge. It was completed in 1905. Its arch, which was constructed from over 5,000 tonnes (4,900 long tons; 5,500 short tons) of stone blocks in just 18 days, is the second largest stone arch in the world, surpassed only by the Friedensbrücke (Syratalviadukt) in Plauen, and the largest railroad stone arch. The arch of the Friedensbrücke, which was built in the same year, has the span of 90 m (300 ft) and crosses the valley of the Syrabach River. The difference between the two is that the Solkan Bridge was built from stone blocks, whereas the Friedensbrücke was built from a mixture of crushed stone and cement mortar. The world's current largest arch bridge is the Chaotianmen Bridge over the Yangtze River with a length of 1,741 m (5,712 ft) and a span of 552 m (1,811 ft). The bridge was open on April 29, 2009 in Chongqing, China.

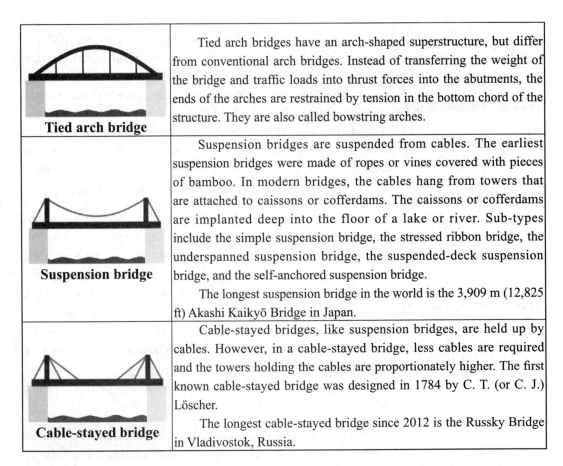

Tied arch bridge	Tied arch bridges have an arch-shaped superstructure, but differ from conventional arch bridges. Instead of transferring the weight of the bridge and traffic loads into thrust forces into the abutments, the ends of the arches are restrained by tension in the bottom chord of the structure. They are also called bowstring arches.
Suspension bridge	Suspension bridges are suspended from cables. The earliest suspension bridges were made of ropes or vines covered with pieces of bamboo. In modern bridges, the cables hang from towers that are attached to caissons or cofferdams. The caissons or cofferdams are implanted deep into the floor of a lake or river. Sub-types include the simple suspension bridge, the stressed ribbon bridge, the underspanned suspension bridge, the suspended-deck suspension bridge, and the self-anchored suspension bridge. The longest suspension bridge in the world is the 3,909 m (12,825 ft) Akashi Kaikyō Bridge in Japan.
Cable-stayed bridge	Cable-stayed bridges, like suspension bridges, are held up by cables. However, in a cable-stayed bridge, less cables are required and the towers holding the cables are proportionately higher. The first known cable-stayed bridge was designed in 1784 by C. T. (or C. J.) Löscher. The longest cable-stayed bridge since 2012 is the Russky Bridge in Vladivostok, Russia.

Bridge types by materials used

The materials used to build the structure are also used to categorize bridges. Until the end of the 18th century, bridges were made out of timber, stone and masonry. Modern bridges are currently built in concrete, steel, fiber reinforced polymers (FRP), stainless steel or combinations of those materials. Living bridges have been constructed of live plants such as tree roots in India and vines in Japan.

Bridge Type	Materials Used
Cantilever	For small footbridges, the cantilevers may be simple beams; however, large cantilever bridges designed to handle road or rail traffic use trusses built from structural steel, or box girders built from prestressed concrete.
Suspension	The cables are usually made of steel cables coated with Zinc, along with most of the bridge, but some bridges are still made of steel reinforced concrete.
Arch	Stone, brick and other such materials that are strong in compression and somewhat so in shear.

Beam	Prestressed concrete is an ideal material for beam bridge construction; the concrete withstands the forces of compression well and the steel rods imbedded within resist the forces of tension. Prestressed concrete also tends to be one of the least expensive materials in construction.
Truss	The triangular pieces of truss bridges are manufactured from straight and steel bars, according to the truss bridge designs.

Words and Expressions

span	[spæn]	n. 时期；跨度；间距　vt. 延续；横跨；贯穿；遍及
terrain	[təˈreɪn]	n. 地带；地形
anchor	[ˈæŋkə(r)]	n. 锚；锚状物；依靠；新闻节目主播；压阵队员　v. 抛锚；停泊；用锚系住；主持节目
log	[lɒg]	n. 原木；（航海、飞行）日志；（船）计程仪　v. 伐木；记入（日志）
plank	[plæŋk]	n. 厚木板；支撑物；政纲条款　vt. 铺板；用放下力
cavern	[ˈkæv(ə)n]	n. 大洞穴　v. 置……于洞穴中；挖空
pier	[pɪə(r)]	n. 码头；桥墩；桥柱；［建］窗间壁；支柱
crossbeam	[ˈkrɔːsˌbiːm]	n. 大梁；横梁
beam	[biːm]	vi. 堆满笑容；照射　vt. 播送；发射；用梁支撑　n. 桁条；光线；（光线的）束；（横）梁
fiber	[ˈfaɪbə]	n. 光纤；（织物的）质地；纤维，纤维物质
compression	[kəmˈpreʃ(ə)n]	n. 压缩，压紧，浓缩，紧缩；加压，压抑；（表现的）简练；应压试验
bending	[ˈbendɪŋ]	n. 弯曲（度），挠度　v. 弯曲
torsion	[ˈtɔːʃn]	n. 扭转；扭曲；［物］［机］扭力
shear	[ʃɪə]	n. 修剪；剪下的东西；大剪刀　v. 剪掉；剪；剥夺
predominate	[prɪˈdɒmɪneɪt]	v. 在……中占优势；支配
suspension	[səˈspenʃ(ə)n]	n. 暂停；中止；悬挂；悬浮液
cable-stayed	[ˈkeɪblˈsteɪd]	n. 斜拉索
truss	[trʌs]	v. 束紧；用桁架支撑　n.（干草的）一捆；一束；构架
arch	[ɑːtʃ]	n. 拱；拱门；拱状物　v. 成拱形；拱起

		adj. 主要的；调皮的
discernible	[dɪ'sɜːnɪbl]	*adj.* 可辨别的
horizontal	[hɒrɪ'zɒnt(ə)l]	*adj.* 水平的；横的 *n.* 水平线；水平面
substructure	['sʌbstrʌktʃə]	*n.* 底部构造；基础；子结构；路基
intermediate	[ˌɪntə'miːdiət]	*adj.* 中间的；中级的 *n.* 中间体；调解人；媒介物 *vi.* 调解；干涉
timber	['tɪmbə]	*n.* 木材；木料
vine	[vaɪn]	*n.* 藤；蔓；攀爬植物
cantilever	['kæntɪliːvə]	*n.* 悬臂；支架 *v.* 由悬臂支撑；如悬臂向外突出
Zinc	[zɪŋk]	*n.* 锌 *vt.* 在……上镀锌
rod	[rɒd]	*n.* 杆；体罚（责打人）用的棍棒
imbedded	[ɪm'bɛdɪd]	*adj.* 嵌入的
triangular	[traɪ'æŋɡjʊlə]	*adj.* 三角形的；以三角形为底的；三方的

Exercises

Part One : Special Terms

1. cantilever bridge _____
2. suspension bridge _____
3. cable-stayed bridge _____
4. 梁桥 _____
5. 桁架桥 _____
6. 拱桥 _____

Part Two : Situational Conversation

(**Peter:** a student   **Li Ya:** a civil engineer)

Peter: Excuse me, can you tell me what you are going to build here?

Li Ya: A multi-storey building.

Peter: A multi-storey building? How many storeys does this building have?

Li Ya: It has 5 storeys or so.

Peter: What do you mean by a multi-storey building? And what is the difference between high-rise building and low-rise one?

Li Ya: According to *the Code for Design of Civil Buildings* (GB 50352-2005), buildings of more than 10 storeys are high-rise, of 7-9 are medium-high, of 4-6 are multi-storeys, and of less than 4 storeys are low-rise.

Peter: Oh, I know it. The classification is only fit for dwellings. How about public buildings?

Li Ya: For public buildings, the low-rise ones are less than 10 meter high, the high-rise ones are more than 24 meters, and the super high ones are over 100 meters.

Peter: I see. What is the predicted height of the building under construction?

Li Ya: Maybe more than twenty meters, but less than twenty-four meters.

Peter: What is the building mainly used for?

Li Ya: Some storeys are for business and the rest for living.

Peter: Which storeys are for department stores or shops?

Li Ya: The lower part, that is to say, the ground and the first floors.

Peter: Can you tell me why?

Li Ya: Because there aren't any partitions in these two storeys. The big rooms are usually for offices and department stores or shops.

Peter: What about rooms in other storeys?

Li Ya: There're many partitions in rooms of the other storeys, so they are mainly used as bedrooms, sitting rooms and bathrooms.

Peter: Right. Is the building Chinese or Western style?

Li Ya: I think it belongs to the Western style. Building with overhanging roofs are usually of the Chinese style, or belong to ancient architecture.

Peter: So the difference between ancient and modern architecture lies in whether there is an overhanging roof, right?

Li Ya: I think so. But there are many other differences in aspects of materials, structures, etc.

Peter: I do not quite understand. Please give me some examples of ancient architecture?

Li Ya: Ok. The Forbidden City (also called Imperial Palaces) in Beijing and the Bell Tower and the Drum Tower in Xi'an, etc., are all glaring examples of ancient architecture.

Peter: And examples of modern architecture?

Li Ya: Too numerous to mention, such as the Great Hall of the People in Beijing and the Oriental Pearl in Shanghai.

Peter: I see. Thank you for your detailed explanation.

Li Ya: It is my pleasure.

Notes:

1. *The Code for Design of Civil Buildings* (GB 50352-2005)

《民用建筑设计通则》为国家标准，编号为 GB 50352—2005，其中解释民用建筑按使用功能可分为居住建筑和公共建筑两大类。民用建筑按地上层数或高度分类划分应符合下列规定。

1）住宅建筑按层数分类：一层至三层为低层住宅，四层至六层为多层住宅，七层至九层为中高层住宅，十层及十层以上为高层住宅。

2）除住宅建筑之外的民用建筑高度不大于24米者为单层和多层建筑，大于24米者为高层建筑(不包括建筑高度大于24米的单层公共建筑)。

3）建筑高度大于100米的民用建筑为超高层建筑。

注：本条建筑层数和建筑高度计算应符合防火规范的有关规定。

2. single-storey building 单层建筑，multi-storey building 多层建筑，high-rise building 高层建筑，low-rise building 低层建筑。

3. partition 隔间。

4. Building with overhanging roofs are usually of the Chinese style, or belong to ancient architecture. 带有悬挑屋顶的建筑通常为中式风格或者古代建筑。

5. The Forbidden City(also called Imperial Palaces) 紫禁城，the Bell Tower，钟楼，the Drum Tower 鼓楼。

Exercise 1: Sentence Patterns

1. How many storeys does this building have? It has 5 storeys or so.

这栋楼有几层？大概有五层。

2. What do you mean by a multi-storey building?

你说的多层建筑是什么意思？

3. What is the difference between high-rise building and low-rise one?

高层建筑与低层建筑的区别是什么？

4. According to *the Code for Design of Civil Buildings* (GB 50352-2005), buildings of more than 10 storeys are high-rise, of 7-9 are medium-high, of 4-6 are multi-storeys, and of less than 4 storeys are low-rise.

根据《民用建筑设计通则》（GB 50352—2005），住宅建筑十层及十层以上为高层住宅，七层至九层为中高层住宅，四层至六层为多层住宅，一层至三层为低层住宅。

5. The classification is only fit for dwellings.

这只适合于住宅建筑的分类。

6. What is the predicted height of the building under construction?

这个正在建设的建筑预计要建多高？

7. Too numerous to mention, such as the Great Hall of the People in Beijing and the Oriental Pearl in Shanghai.

举不胜举，比如北京的人民大会堂，还有上海的东方明珠电视塔。

Exercise2: Complete the Following Dialogue in English

(A: a student majoring in civil engineering   B: a teacher)

A: Can you tell me how to classify the buildings according to usage?

B: _____.

（一般说来，建筑物按照使用情况可分为三大类，即：民用建筑、工业建筑和农业建筑。）

A: What are civil buildings?

B: _____.

（民用建筑包括居住建筑和公共建筑。）

A: Can you list some examples of public buildings?

B: _____.

（举不胜举。例如写字楼、酒店、商店、学校、机场、车站等都属于公共建筑。）

A: I see. Can I say the National Stadium is one of the most glaring examples of buildings?

B: _____.

（当然可以，国家体育馆不仅壮观，而且坚固，其主体结构设计使用年限达到100年。）

A: Is there anything special in the design of its main structure?

B: _____.

（是的，基于大跨度空间的需要，屋顶采用钢结构，其他主要承重构件用钢筋混凝土结构。）

A: Is the National Stadium a high-rise building?

B: _____.

（是的，对于公共建筑而言，建筑物两层以上高度 24 m 以上即为高层建筑。）

A: What about the high-rise for residential buildings?

B: _____.

（十层以上的住宅建筑才算是高层。）

A: Oh, I see.

Part Three: Reading Comprehension

When students arrive at campus with their parents, both parties often assume that the school will function in loco parentis, watching over its young charges, providing assistance when needed. Colleges and universities present themselves as supportive learning communities —as extend families, in a way. And indeed, for many students they become a home away from home. This is why graduates often use another Latin term, *alma mater*, meaning "nourishing mother". Ideally, the school nurtures its students, guiding them toward adulthood. Lifelong friendships are formed, teachers become mentors（导师）, and the academic experience is complemented by rich social interaction. For some students, however, the picture is less rosy. For a significant number, the challenges can be become overwhelming.

In reality, administrator at America colleges and universities are often obliged to focus as much on the generation of revenue as on the new generation of students. A trouble or even severely disturbed student can easily fall through the cracks. Public institutions in particular are often faced with tough choices about which student support services to fund, and how to manage such things as soaring health-care cost for faculty and staff. Private schools are feeling the pinch as well. Ironically, although tuition and fees can increase as much as 6.6 percent in a single year, the high cost of doing business at public and private institutions means that students are not necessarily receiving more support in return for increased tuition and fees. To compound the problem, students may be reluctant to seek help even when they desperately need it.

Unfortunately, higher education is sometimes more of an information delivery system than a responsive, collaborative process. Just as colleges are sometimes ill-equipped to respond to the challenges being posed by today's students, so students themselves are sometimes ill-equipped to respond to the challenges posed by college life. Although they arrive at campus with high expectations, some students struggle with chronic shyness, learning disabilities, addiction, or eating disorders. Still others suffer from acute loneliness, mental illness, or even rage.

We have created cities of youth in which students can pass through unnoticed, their voices rarely heard, their faces rarely seen. As class size grows in response to budget cuts, it becomes even less likely that troubled students, or even severely disturbed students, will be noticed. When they're not, the results can be tragic.

1. How do students feel about college and universities?

A) admiring B) disappointed C) indifferent D) affectionate

2. What's the ideal image of colleges and universities ?

   A) They are places where academic requirements are loose.

   B) They are places where students can have colorful social experience.

   C) They nurture students and guide them to grow into adults.

   D) They teach students how to spend their youth time best.

3. Why do American colleges and universities often neglect troubled or even severely disturbed students ?

   A) They view academic management as their only task.

   B) They don't have so much energy or money to focus on the needs of students.

   C) They don't care about the students' psychological health.

   D) They focus mainly on improving the salaries for faculty and staff.

4. Which of the following information can be got from the third paragraph ?

   A) Private schools in American never feel that they are short of money.

   B) American students receive more support as tuition and fees increase.

   C) American colleges and universities fail to respond to and help students in time sometimes.

   D) American college students seek help only when they suffer from serious psychological illness.

5. What's the author's attitude towards American higher education?

   A) critical   B) neutral   C) praising   D) unintended

Part Four: Translation Skills

翻译技巧：转换法

转换法指翻译过程中为了使译文符合目标语的表述方式、方法和习惯而对原句中的词类、句型和语态等进行转换。具体来说，就是在词性方面，把名词转换为代词、形容词、动词；把动词转换成名词、形容词、副词、介词；把形容词转换成副词和短语。在句子成分方面，把主语转换成状语、定语、宾语、表语；把谓语转换成主语、定语、表语；把定语转换成状语、主语；把宾语转换成主语。在句型方面，把并列句转换成复合句，把复合句转换成并列句，把状语从句转换成定语从句。在语态方面，可以把主动语态转换为被动语态。如：

例1： The reform and opening policy is supported by the whole Chinese people.

翻译：改革开放政策受到了全中国人民的拥护。（动词转名词）

详解：在本句的翻译中，原文中的谓语动词结构 is supported by 在翻译中被转换成了名词结构"拥护"。这样的翻译转换处理是由于在汉语中不太习惯用被动意义来表达对于政府的支持，而"受到全国人民的拥护"更符合汉语的表达习惯，更有利于读者的理解。

例2： In his article the author is critical of man's negligence toward his environment.

翻译：作者在文章中，对人类疏忽自身环境作了批评。（形容词转名词）

详解：本句的翻译中，原文中的形容词结构 critical of 被转化为了名词结构"批评"。之所以这样处理，是因为在汉语中，对于批评这个意义主要限定于动词和名词词性，因此，将这个形容词结构转化为名词结构，更符合汉语表达习惯。

例3： In some of the European countries, the people are given the biggest social benefits such as medical insurance.

翻译：在有些欧洲国家里，人民享受最广泛的社会福利，如医疗保险等。（被动语态转主动语态）

详解：在中文里，"给予"这个词通常用作主动含义，而被动意义很少使用，也不太符合中文的表达习惯，另外该句英文中"被给予最大的社会福利"从意义上来说，表达的就是"人民能享受到最广泛的社会福利"。因此在翻译时，将 are given the biggest social benefits 转换为了主动语态。

Translation Exercise 1

结构钢与钢筋混凝土的使用使传统的施工作业发生了明显的变化。多层建筑再也没必要采用厚的石墙或砖墙，且防火地面施工变得更容易。这些变化有利于降低建筑的成本。它也使建造高度更高和跨度更大的建筑物成为可能。

Translation Exercise 2

建筑的设计和施工是由称为建筑规范的市政细则来控制的。这些规范用以保护公众的健康和安全。每个城市和城镇允许编制或采用其自己的建筑规范，并且在该城市或城镇只有特定的规范才有合法的地位。由于建筑规范编制的复杂性，美国的城市通常将它们的建筑规范以三个典型规范中的一个为基础，即统一建筑规范、标准建筑规范或基本建筑规范。这些规范包括诸如使用和居住的要求、防火要求、供热和通风要求以及结构的设计。

Unit 5　Building Code

Learning Objectives

After completing this unit, you will be able to do the following:
- √ Grasp the main idea and the structure of the text;
- √ Master the key language points and grammatical structure in the text;
- √ Understand the basic knowledge of building code;
- √ Conduct a series of reading, listening, speaking and writing activities related to the theme of the unit.

Outline

The following are the main sections in this unit.
1. Warm-up Activity
2. Text
3. Words and Expressions
4. Exercises

Terms of Building Code

In this unit, you will learn the meanings of terms listed below.
building code
workmanship
electrical system

Vocabulary

Listed below are some words appearing in this unit that you should make them a part of your vocabulary.
alteration
accomplish
performance
sewer
install
sewage

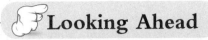 Looking Ahead

Building Code

The 2012 International Building Code (IBC) covers all buildings except detached one and two family dwellings and townhouses not more than three storeys in height. This comprehensive code features time-tested safety concepts, structural, and fire and life safety provisions covering means of egress, interior finish requirements, comprehensive roof provisions, seismic engineering provisions, innovative construction technology, occupancy classifications, and the latest industry standards in material design. It is founded on broad-based principles that make the use of new materials and new building designs possible.

Introduction

With the development of society, building codes are constantly renewed. The following are the two developing stages of building codes.

Originally, building codes were of the specifications type. They require that all construction be accomplished using specified materials in a specified way. The builder has very little choice in the material or the methods of construction.

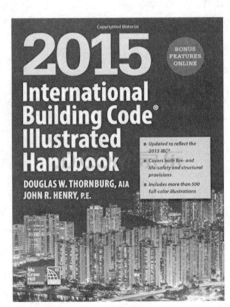

But times changed and new materials were developed. Since World War II there has been a significant change toward building codes of the performance type. In this type of code, the performance standards of a material or structure are outlined. The builder is free to select materials or building techniques which will meet these standards.

Warm-up Activity

Do you know what a building code is?

 Text

Building Code

A building code is a set of detailed regulations to ensure that all the buildings meet certain minimum standards of health and safety. Building codes have been enacted to protect citizens from any harm likely to come to them because of unhealthy or unsafe conditions.

The plan for all new constructions must be approved by officials of the buildings departments before construction begins. They must be able to inspect all equipments, materials and workmanship before the building is approved for occupancy. If the equipments, workmanship, or materials do not meet the standards of the buildings, these officials have the right to order that the necessary changes be made.

Occasionally, the owner of a building might want to make a basic alteration in the electrical, heating, or plumbing systems. He may also want to make a basic change in the structure. In this case, the building department must approve the alterations beforehand. After the alterations have been completed, the officials must once again inspect the workmanship and materials.

Originally, building codes were of the specifications type. They require that all construction be accomplished using specified materials in a specified way. The builder have very little choice in the material or the methods of construction. But times changed and new materials were developed. Since World War II there has been a significant change toward building codes of the performance type. In this type of code, the performance standards of a material or structure are outlined. The builder is free to select materials or building techniques which will meet these standards.

For example, a specification-type code for a house sewer will simply specify that cast iron pipe of a certain quality and size be used. It will also specify that the pipe be installed in a specified manner. The plumbing contractor has no choice or say in the matter. In a performance-type building code, however, the code will specify that the piping not be affected by any corrosive or harmful substances in the sewage or in the soil in which the pipe is buried. The code will also require that the pipe meet certain minimum strength requirements. In addition, the code will specify that the pipe not be affected by temperature changes within a specified range. The plumbing contractor is free to use plastic pipe, cast-iron pipe, or gold pipe if he wants to, as long as he can show that the pipe does in fact meet the specified standards.

 Words and Expressions

minimum	['mɪnɪməm]	n. 最小量
		adj. 最小的
enact	[ɪ'nækt]	vt. 颁布
performance	[pə'fɔ:m(ə)ns]	n. 执行；性能
sewer	[su:ə]	n. 排水管，污水渠
install	[ɪn'stɔ:l]	vt. 装配，安装
affect	[ə'fekt]	vt. 影响
sewage	['su:ɪdʒ]	n. 污水，下水管（系统）
range	[reɪn(d)ʒ]	n. 范围，射程
in this case		在这种情况下
in addition		另外，还有

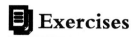 **Exercises**

Part One: Special Terms

1. the building department   _____
2. sewer   _____
3. cast iron pipe   _____
4. 建筑规范   _____
5. 工艺   _____
6. 电力规范   _____

Part Two : Situational Conversation

A: One month ago, the Main Contractor issued a Site Safety Program to all the subcontractors on site. I hope that each subcontractor has prepared his own Safety Scheme according to that Site Safety Program. Now I'd like to be informed of your Safety Scheme.

B: OK, here is our Safety Scheme. In compliance with the Site Safety Program, it details all the precautions and procedures for structure construction.

A: Do you have a group of professionals responsible for implementing this Safety Scheme?

B: Yes. From the beginning of this project, the Safety Manager was appointed, and he has three assistants respectively responsible for the safety of each section of the site.

A: Good. Now let's go to the site and check if all the safety regulations are strictly observed. In order to appraise your performance on site safety, we will record all the non-compliance found during our safety patrol. Please put on your helmets before you step out of the meeting room.

B: All right, this way, please.

A: I'd like to speak to this welder first. Did you have any training courses for safety education?

C: Yes. I learned all the safety regulations related to welding in the class.

A: It's fine that you wear the helmet and protective goggles, but where are your hard shoes? Didn't your manager give one pair to you?

C: Yes, he did. I'm sorry. I forgot to wear them today.

A: Don't forget again. Did you get a "Work Permit" for the welding here?

C: Sure, here it is.

Notes：

1. Do you have a group of professionals responsible for implementing this Safety Scheme? 你们有没有成立专家组负责这个安全计划的实施？ "responsible for implementing this safety scheme" 是形容词短语作后置定语。

2. I learned all the safety regulations related to welding in the class. 我在培训班上学到了所有焊接方面的安全规章。 "related to welding in the class" 是过去分词短语作定语。

Exercise 1: Sentence Patterns

1. Now I'd like to be informed of your Safety Scheme. 现在我想听取你们的安全方案。

2. In compliance with the Site Safety Program, it details all the precautions and procedures for structure construction. 按照现场安全程序，本方案详细说明了结构施工的所有预防措施和步骤。

3. Please put on your helmets before you step out of the meeting room. 出会议室之前，请带上安全帽。

4. I'd like to speak to this welder first. 我想先找焊接工聊聊。

5. It's fine that you wear the helmet and protective goggles, but where are your hard shoes? 你戴了安全帽和护目镜是对的，但是你的硬底鞋在哪儿？

Exercise 2: Complete the Following Dialogue in English

(A: a supervisor B: a welder)

A: _____?

（你上过安全教育培训课吗？）

B: Yes. I learned all the safety regulations related to welding in the class.

A: _____
_____?

（你戴了安全帽和护目镜是对的，但是你的硬底鞋在哪儿？你们经理没有给你一双吗？）

B: Yes, he did. I'm sorry. I forgot to wear them today.

A: _____

（别再忘了。你有焊接的工作许可证吗？）

B: Sure, here it is.

Part Three: Reading Comprehension

Despite the brouhaha (骚动) over stolen e-mails from the University of East Anglia, the science of climate change is well enough established by now that we can move on the essential question: what's the damage going to be?

"The total bill, if emissions are left unchecked, could reach 20 percent of annual per capita income", says Nicholas Stern, the British economist who led an influential Whitehall-sponsored study. William Nordhaus, a Yale economist, puts his "best guess" at 2.5 percent of yearly global GDP. And according to Dutch economist Richard Tol, the economic impact of a century's worth of climate change is "relatively small" and "comparable to the impact of one or two years of economic growth."

These estimates aren't just different— they're different by an order of magnitude. And while some might dismiss the cost estimates as mere intellectual exercises, they're intellectual exercises with real impact. The Copenhagen meeting may be a bust, but countries from the United States to China are individually considering cap-and-trade schemes, carbon taxes, and other policies aimed at curtailing greenhouse gases. To be effective, a tax or cap-and-trade charge would have to force today's emitters to pay the true "social cost of carbon" — in other words, the amount of damage a ton of carbon will cause in the coming centuries.

Figuring out what that cost is, however, is no simple task. That's largely because most of the bill won't come due for many decades. A ton of carbon dioxide emitted today will linger in the air for one to five centuries. Virtually every cost study shows that, even if economic growth continues apace (快速地) and there's no effort to slash emissions, the damage from climate change will be negligible until at least 2075. It could take 100 years before we see noticeably negative effects, and even more before we need to launch massive construction projects to mitigate (减轻) the damage.

1. What can we learn from the first paragraph?

   A) Those stolen e-mails should not be made a fuss of.

   B) It is time to talk about damage caused by climate change.

   C) The science of climate should have been established.

   D) Climate change is essential to human begins.

2. What's the damage brought by climate change according to the author?

   A) It account for 20% of annual per capita income.

   B) It is about 2.5% of the yearly global GDP.

   C) Its damage is overestimated by many economists.

D) Economists haven't reached consensus yet.
3. What should be done to reduce carbon emissions?
   A) Gas emitters should pay for the damage.
   B) Policies should aim at reducing carbon emission.
   C) Social cost of carbon should be shown to the public.
   D) The damage should not be neglected.
4. Why is it hard to figure out the social cost of carbon?
   A) Too many factors should be taken into account.
   B) There is no effort aiming at carbon reduction.
   C) The damage cannot be seen until years later.
   D) The damage will last for years before eliminated.
5. What can we learn about the present climate change?
   A) The damage is somewhat exaggerated.
   B) Action will be taken the moment people realize it.
   C) Measures should be taken immediately to tackle it.
   D) The negative effect will not be significant for a century.

Part Four: Translation Skills

拆句法和合并法

拆句法和合并法是两种相对应的翻译方法。拆句法是把一个长而复杂的句子拆译成若干个较短、较简单的句子，通常用于英译汉；合并法是把若干个短句合并成一个长句，一般用于汉译英。汉语强调意合，结构较松散，因此简单句较多；英语强调形合，结构较严密，因此长句较多。所以汉译英时要根据需要注意利用连词、分词、介词、不定式、定语从句、独立结构等把汉语短句连成长句；而英译汉时又常常要在原句的关系代词、关系副词、主谓连接处、并列或转折连接处、后续成分与主体的连接处以及意群结束处将长句切断，译成汉语分句。这样就可以基本保留英语语序，顺译全句，顺应现代汉语长短句交替、单复句相间的句法修辞原则。

例1：Increased cooperation with China is in the interests of the United States.

翻译：同中国加强合作，符合美国的利益。（在主谓连接处拆译）

详解：根据意义上的需要，一个复杂的句子可以拆分为两个小句。increased 的词根（increase）本就有动词词性的意思，所以把 increased cooperation with China，翻译成与中国加强合作合乎情理，句子也短而精炼，be in the interests of 表示与"与……利益相符"，把 be 翻译成"符合"，而非"是"，符合汉语的词组习惯。

例 2： I wish to thank you for the incomparable hospitality for which the Chinese people are justly famous throughout the world.

翻译：我要感谢你们无与伦比的盛情款待。中国人民正是因为这种热情好客而闻名世界的。（在定语从句前拆译）

详解：从 for which 这个结构可以看出后半句 the Chinese people are justly famous throughout the world 是一个定义从句。定义从句是一个复杂句，此时就可以用拆句法，把定语从句翻译成一个小句，其主句也可以翻译成一个小句。通过"for which"可以判断两个小句的关系是表原因，所以前一句翻译成"我要感谢你们无与伦比的盛情款待"，后一小句翻译成"中国人民正是因为这种热情好客而闻名于世的"，符合汉语语言规范。

例 3： 中国是一个发展中的沿海大国。中国高度重视海洋的开发和保护，把发展海洋事业作为国家发展战略。

翻 译： As a major developing country with a long coastline, China attaches great importance to marine development and protection, and takes it as the state's development strategy.（合并法）

详解：此处原句被翻译成一个长长的英语单句。原句的第一句被译成一个由 as 引导的介词短语，表示作为一个发展中的沿海大国，主语仍然是中国，主管两个并列动词："高度重视"和"把……作为"，其对应的英文词组就是 attach importance to 和 take or regard ...as，符合英语的惯用语，也表达了汉语的基本意思。

Translation Exercise 1

建筑规范是一套详细的条例，目的在于确保所有建筑物符合一定的卫生和安全的标准。建筑规范的颁布，是为了保护其公民不致因不卫生或不安全的条件受到任何可能出现的危害。

Translation Exercise 2

所有新建工程的计划必须在施工开始前得到建筑部门的批准。他们必须在该建筑被批准占用前对所有设备、材料和工艺进行检查。如果设备、工艺或材料不符合建筑规范的标准，建筑部门有权命令其进行必要的整改。

Unit 6 Construction Cost Estimate

Learning Objectives

After completing this unit, you will be able to do the following:
- √ Grasp the main idea and the structure of the text;
- √ Master the key language points and grammatical structure in the text;
- √ Understand the types of construction cost estimate;
- √ Conduct a series of reading, listening, speaking and writing activities related to the theme of the unit.

Outline

The following are the main sections in this unit.
1. Warm-up Activity
2. Text
3. Words and Expressions
4. Exercises

Terms of Construction Cost Estimate

In this unit, you will learn the meanings of terms listed below.
subcontractor
cantilever
assess
perspective

Vocabulary

Listed below are some words appearing in this unit that you should make them a part of your vocabulary.
substantial
institutional
categories
makeup
designate
overhead

Looking Ahead

Cost estimating is one of the most important steps in project management. A cost estimate establishes the base line of the project cost at different stages of development of the project. A cost estimate at a given stage of project development represents a prediction provided by the cost engineer or estimator on the basis of available data.

Introduction

Construction cost estimates may be viewed from different perspectives because of different institutional requirements. In spite of the many types of cost estimates used at different stages of a project, cost estimates can best be classified into three major categories according to their functions. A construction cost estimate serves one of the three basic functions: design, bid and control.

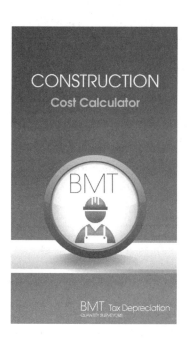

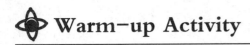

Warm-up Activity

Could you talk the importance of cost estimates in project management?

Construction Cost Estimate

Cost estimating is one of the most important steps in project management. A cost estimate establishes the base line of the project cost at different stages of development of the project. A cost estimate at a given stage of project development represents a prediction provided by the cost engineer or estimator on the basis of available data. According to the American Association of Cost Engineers, cost engineering is defined as that area of engineering practice where engineering judgment and experience are utilized in the application of scientific principles and techniques to the problem of cost estimation, cost control and profitability. Virtually all cost estimates are performed according to one or some combination of the following basic approaches.

Production function

In microeconomics, the relationship between the output of a process and the necessary resources is referred to as the production function. In construction, the production function may be expressed by the relationship between the volume of construction and a factor of production such as labor or capital. A production function relates the amount or volume of output to the various inputs of labor, material and equipment. For example, the amount of output Q may be derived as a function of various input factors $X_1, X_2, ..., X_n$ by means of mathematical and/or statistical methods. Thus, for a specified level of output, we may attempt to find a set of values for the input factors so as to minimize the production cost. The relationship between the sizes of a building project (expressed in square foot) to the input labor (expressed in labor hours per square foot) is an example of a production function for construction.

Empirical cost inference

Empirical estimation of cost functions requires statistical techniques which relate the cost of constructing or operating a facility to a few important characteristics or attributes of the system. The role of statistical inference is to estimate the best parameter values or constants in an assumed cost function. Usually, this is accomplished by means of regression analysis techniques.

Unit costs for bill of quantities

A unit cost is assigned to each of the facility components or tasks as represented by the bill of quantities. The total cost is the summation of the products of the quantities multiplied by the corresponding unit costs. The unit cost method is straightforward in principle but quite laborious in application. The initial step is to break down or disaggregate a process into a number of tasks. Collectively, these tasks must be completed for the construction of a facility. Once these tasks are defined and quantities representing these tasks are assessed, a unit cost is assigned to each and then the total cost is determined by summing the costs incurred in each task. The level of detail in decomposing into tasks will vary considerably from one estimate to another.

Allocation of joint costs

Allocations of cost from existing accounts may be used to develop a cost function of an operation. The basic idea in this method is that each expenditure item can be assigned to particular characteristics of the operation. Ideally, the allocation of joint costs should be causally related to the category of basic costs in an allocation process. In many instances, however, a causal relationship between the allocation factor and the cost item cannot be identified or may not exist. For example, in construction projects, the accounts for basic costs may be classified according to: labor, material, construction equipment, construction supervision, and general office overhead. These basic costs may then be allocated proportionally to various tasks which are subdivisions of a project.

Construction cost constitutes only a fraction, though a substantial fraction, of the total project cost. However, it is the part of the cost under the control of the construction project manager. The required levels of accuracy of construction cost estimates vary at different stages of project development, ranging from ball park figures in the early stage to fairly reliable figures for budget control prior to construction. Since design decisions made at the beginning stage of a project life cycle are more tentative than those made at a later stage, the cost estimates made at the earlier stage are expected to be less accurate. Generally, the accuracy of a cost estimate will reflect the information available at the time of estimation.

Construction cost estimates may be viewed from different perspectives because of different institutional requirements. In spite of the many types of cost estimates used at different stages of a project, cost estimates can best be classified into three major categories according to their functions. A construction cost estimate serves one of the three basic functions: design, bid and control. For establishing the financing of a project, either a design estimate or a bid estimate is used.

Design estimates For the owner or its designated design professionals, the types of cost estimates encountered run parallel with the planning and design as follows:

(1) Screening estimates (or order of magnitude estimates);

(2) Preliminary estimates(or conceptual estimates);

(3) Detail estimates (or definitive estimates);

(4) Engineer's estimates based on plans and specifications.

For each of these different estimates, the amount of design information available typically increase.

Bid estimates For the contractor, a bid estimate submitted to the owner either for competitive bidding or negotiation consists of direct construction cost including field supervision, plus a makeup to cover general overhead and profits. The direct cost of construction for bid estimates usually derived from a combination of the following approaches.

Control estimates For monitoring the project during construction, a control estimate is derived from available information to establish:

Budget estimate for financing;

Budgeted cost after contracting but prior to construction;

Estimated cost to completion during the progress of construction;

Design estimates

In the planning and design stages of a project, various design estimates reflect the progress of the design. At the very early stage, the screening estimate or order of magnitude estimate is usually made before the facility is designed, and must therefore rely on the cost data of similar facilities built in the past. A preliminary estimate or conceptual estimate is based on the conceptual design of the facility at the state when the basic technologies for the design are known.The detailed estimate or definitive estimate is made when the scope of work is clearly defined and the detailed design is in progress so that the essential features of the facility are identifiable. The engineer's estimate is based on the completed plans and specifications when they are ready for the owner to solicit bids from construction contractors. In preparing these estimates, the design professional will include expected amounts for contractors' overhead and profits.

The costs associated with a facility may be decomposed into a hierarchy of levels that are appropriate for the purpose of cost estimation. The level of detail in decomposing the facility into tasks depends on the type of cost estimate to be prepared. For conceptual estimates, for example, the level of detail in defining tasks is quite coarse; for detailed estimates, the level of detail can be quite fine.

As an example, We consider the cost estimates for a proposed bridge across a river. A screening estimate is made for each of the potential alternatives, such as a tied arch bridge or a cantilever truss bridge. As the bridge type is selected, e.g. the technology is chosen to be a tied arch bridge instead of some new bridge form; a preliminary estimate is made on the basis of the layout of the selected bridge form on the basis of the preliminary or conceptual design.When the detailed design has progressed to a point when the essential details are known, a detailed estimate is made on the basis of the well-defined scope of the project. When the detailed plans and specifications are completed, an engineer's estimate can be made on the basis of items and quantities of work.

Bid estimates

The contractor's bid estimates often reflect the desire of the contractor to secure the job as well as the estimating tools at its disposal. Some contractors have well established cost estimating procedures while others do not. Since only the lowest bidder will be the winner of the contract in most bidding contests, any effort devoted to cost estimating is a loss to the contractor who is not a successful bidder. Consequently, the contractor may put in the least amount of possible effort for making a cost estimate if it believes that its chance of success is not high.

If a general contractor intends to use subcontractors in the construction of a facility, it may solicit price quotations for various tasks to be subcontracted to specialty subcontractors. Thus, the general subcontractor will shift the burden of cost estimating to subcontractors. If all or part of the construction is to be undertaken by the general contractor, a bid estimate may be prepared on the basis of the quantity takeoffs from the plans provided by the owner or on the basis of the construction procedures devised by the contractor for implementing the project. For example, the cost of a footing of a certain type and size may be found in commercial publications on cost data which can be used to facilitate cost estimates from quantity takeoffs. However, the contractor may want to assess the actual cost of construction by considering the actual construction procedures to be used and the associated costs if the project is deemed to be different from typical designs. Hence, items such as labor, material and equipment needed to perform various tasks may be used as parameters for the cost estimates.

Control estimates

Both the owner and the contractor must adopt some base line for cost control during the construction. For the owner, a budget estimate must be adopted early enough for planning long term financing of the facility. Consequently, the detailed estimate is often used as the budget estimate since it is sufficient definitive to reflect the project scope and is available long before the engineer's estimate. As the work progresses, the budgeted cost must be revised periodically to reflect the estimated cost to completion. A revised estimated cost is necessary either because of change orders initiated by the owner or due to unexpected cost overruns or savings.

For the contractor, the bid estimate is usually regarded as the budget estimate, which will be used for control purposes as well as for planning construction financing. The budgeted cost should also be updated periodically to reflect the estimated cost to completion as well as to insure adequate cash flows for the completion of the project.

 Words and Expressions

empirical	[em'pɪrɪk(ə)l]	*adj.* 完全根据经验的
regression	[rɪ'greʃ(ə)n]	*n.* 回归，复原
straightforward	[streɪt'fɔːwəd]	*adj.* 坦率的，简单的，易懂的
initial	[ɪ'nɪʃəl]	*adj.* 最初的，初始的
disaggregate	[dɪs'ægrɪgeɪt]	*vt.* 使崩溃，分解，解开聚集
collective	[kə'lektɪv]	*adj.* 全体的，共同的

overhead	[ˌəʊvə'hed]	n. 企业一般管理项目费，杂项开支
substantial	[səb'stænʃ(ə)l]	adj. 大量的，真实的，充实的
perspective	[pə'spektɪv]	n. 透视图；远景，前途
institutional	[ɪnstɪ'tjuːʃ(ə)n(ə)l]	adj. 制度上的，惯例的，学会的
category	['kætɪɡ(ə)rɪ]	vt. 加以类别，分类
designate	['dezɪɡneɪt]	vt. 指明，指出，指派
parallel	['pærəlel]	adj. 平行的，类似的
		n. 平行线
makeup	[meɪkʌp]	n. 构造
identifiable	[aɪ'dentɪfæɪəb(ə)l]	adj. 可以确认的
solicit	[sə'lɪsɪt]	vt. 恳求
hierarchy	['haɪərɑːki]	n. 层次，层级
truss	[trʌs]	vt. 捆绑
deem	[diːm]	vt. 认为，相信
assess	[ə'ses]	vt. 估定，评定
revise	[rɪ'vaɪz]	vt. 修订，校订
takeoff	['teɪkɒf]	n. 估计量
ball park		（数量，质量或程度）相近，相似
a set of		一组，一套
in principle		原则上，大体上
break down		毁掉，制服，压倒
run parallel with		与……平行
as follows		如下
in progress		前进，进行中
be ready to		预备，即将
be appropriate for		对……合适
at one's disposal		随某人自由处理，由某人随意支配

Exercises

Part one: Special Terms

1. be referred to as　　＿＿＿＿＿＿＿＿＿＿＿＿＿＿＿＿
2. be derived from　　＿＿＿＿＿＿＿＿＿＿＿＿＿＿＿＿

3. run parallel with　　＿＿＿＿＿＿＿＿＿＿＿＿＿＿＿＿＿

4. 工程费用分摊　　　＿＿＿＿＿＿＿＿＿＿＿＿＿＿＿＿＿

5. 预算成本　　　　　＿＿＿＿＿＿＿＿＿＿＿＿＿＿＿＿＿

6. 现金流转　　　　　＿＿＿＿＿＿＿＿＿＿＿＿＿＿＿＿＿

Part Two: Situational Conversation

Mr. Wang: I know a cost estimate is one of the most important steps in project management. Can you tell me something about it?

Mr. Zhao: OK. The importance of cost estimating lies in the fact that it establishes the base line of the project cost at different stages of the project development. And a cost estimate at a given stage of project development represents a prediction provided by the cost engineer or estimator on the basis of available data.

Mr. Wang: How is the cost estimation performed? In other words, what are the basic approaches adopted to estimate the construction cost?

Mr. Zhao: Virtually, all cost estimations are performed according to one or some combination of the following basic approaches: production function, emprical cost inference, unit costs for bill of quantities, allocation of joint costs.

Mr. Wang: What does production function mean in construction?

Mr. Zhao: In construction, the production function refers to the relationship between the construction volumes and production factors such as labor or capital. It relates the output amount or volume to the various inputs of labor, material and equipment.

Mr. Wang: Can you give me an example?

Mr. Zhao: The relationship between the size of a building project (expressed in square foot) to the input labor (expressed in labor hours per square foot) is a typical example of the production function in construction.

Mr. Wang: I understand it now. How is the empirical cost inferred?

Mr. Zhao: In the empirical cost inference, the regression analysis techniques are required, the role of which is to estimate the best parameter values or constants in an assumed cost function.

Mr. Wang: Is the unit cost method the simplest one among the four approaches? Does it mean the total cost is the summation of the product quantities multiplied by the corresponding unit cost?

Mr. Zhao: Yes, you are right. The unit cost method is straightforward in principle but quite laborious in application. First the whole construction process is disaggregated into a number of tasks. Once these tasks are defined and quantities representing these tasks are assessed, a unit cost is assigned to each and then the total cost is determined by summing the cost incurred in each task. The level of detail in decomposing into tasks will vary considerably from one estimate to another.

Mr. Wang: What is the category of basic costs in an allocation process?

Mr. Zhao: In construction projects, the accounts for basic costs can be classified according to labor, material, construction supervision and general office overhead. These basic costs are then allocated proportionally to various tasks which are subdivisions of a project.

Notes:

1. Cost estimating 造价估算。在工程项目的不同阶段，项目投资有估算、预算、结算和决算等不同称呼，这些"算"的依据和作用不同，其准确性也"逐渐明晰"，一个比一个更真实地反映项目的实际投资。

2. What are the basis approaches adopted to estimate the construction cost? 建筑成本估算采用哪些基本方法？其中 adopted to estimate the construction cost 作定语，修饰 approaches。adopt 采用，采纳；adapt 适应，改编。

3. empirical cost inference 成本经验推断。

4. unit costs for bill of quantities 工程量清单单价。

5. allocation of joint costs 联合费用的分摊。

6. the regression analysis techniques 回归分析法。

7. the best parameter values or constants 最佳参数值或常数。

8. general office overhead 普通的办公管理费。

Exercise 1: Sentence Patterns

1. The importance of cost estimating lies in the fact that it establishes the base line of the project cost at different stages of the project development. 造价估算的重要性在于它设立了项目发展不同阶段项目成本的底线。

2. Virtually, all cost estimates are performed according to one or some combination of the following basic approaches. 事实上，所有的成本估算都是根据如下基本方法中的一种或多种相结合来进行的。

3. In construction, the production function refers to the relationship between the construction volumes and production factors such as labor or capital. 在建筑中，生产函数表示施工量与生产因素之间的关系，如劳动力或资金之间的关系。

4. The total cost is the summation of the product quantities multiplied by the corresponding unit cost. 总造价为工程量与相应单价的乘积之和。

5. The unit cost method is straightforward in principle but quite laborious in application. 单价法原则上简单易懂，但应用时相当烦琐。

6. The level of detail in decomposing into tasks will vary considerably from one estimate to another. 分解作业的细分程度对于不同的估价项目有很大的区别。

7. In construction projects, the accounts for basic costs can be classified according to labor, material, construction supervision and general office overhead.
在施工项目中，可根据劳动力、材料、施工监理和普通的办公管理费对基本成本的账目进行分类。

Exercise 2: Complete the Following Dialogue in English

A: What is the main purpose of control estimates?

B: _____.

（为了在施工期间对项目进行监督，根据可利用资料作出造价控制从而确定融资概算、签约后施工前的预算成本以及施工过程中完成部分的预算价值。）

A: Both the owner and the contractor must adopt some basic line for cost control during the construction, am I right?

B: _____.

（是的。对于业主而言，为了做项目的长期投资计划，必须尽可能地采用概算。）

A: Is it necessary for the owner to revise the budgeted cost?

B: _____.

（是的，无论是由于业主变更订单或由于意外费用超支或节约，修正预算成本都是非常必要的。）

A: How about the control estimates for the contractor?

B: _____.

（对于承包商而言，投标报价通常被认为是概算，用于控制投资的目的以及规划建设资金。）

A: Should the budgeted cost be also updated periodically?

B: _____.

（是的。概算定期修正的目的是反映完成的预算价值以及确保项目完成有足够的现金流转。）

Part Three: Reading Comprehension

Questions 1 to 5 are based on the following passage.

British psychologists have found evidence of a link between excessive Internet use and depression, a research has shown.

Leeds University researchers, writing in the *Psychopathology* journal, said a small proportion of Internet users were classed as Internet addicts and that people in this group were more likely to be depressed than non-addicted users.

The article on the relationship between excessive Internet use and depression, a questionnaire-based study of 1,319 young people and adult, used date gathered from respondents to links placed on UK-based social networking sites.

The respondents answered questions about how much time they spent on the Internet and what they used it for; they also completed the Beck Depression Inventory—a series of questions designed to measure the severity of depression.

The six-page report, by the university's Institute of Psychological Sciences, said 18 of the people who completed the questionnaire were Internet addicts. "Our research indicates that excessive Internet use is associated with depression, but what we don't know is which comes first—are depressed people drawn to the Internet cause depression?"

The article's lead author, Dr. Catriona Morrison said, "What is clear is that, for a small part of people, excessive use of the Internet could be a warning signal for depressive tendencies."

The age range of all respondents was between 16 and 51 years, with a mean age of 21.24. The mean age of the 18 Internet addicts, 13 of whom are male and 5 females, was 18.3 years. By comparing the scale of depression within this group to that within a group of 18 non-addicted Internet users, researchers found the Internet addicts had a higher incidence of moderate to severe depression than non-addicts. They also discovered that addicts spent proportionately more time browsing sexually pleasing websites, online gaming sites and online communities.

"This study reinforces the public speculation（推测）that over-engaging in websites that serve to replace normal social function might be linked to psychological disorders like depression and addiction," Morrison said. "We now need to consider the wider social implications of this relationship and establish clearly the effects of excessive Internet use on mental health."

1. Internet addicts are people who _____
    A) use the Internet more than enough   B) feel depressed when using the Internet
    C) seldom connect to the Internet   D) feel depressed without the Internet

2. What was collected as data by the researchers?
    A) The number of users of UK-based social networking sites.
    B) Respondents' visits to UK-based social networking sites.

C) Links on UK-based social networking sites.

D) Respondents' answers to a questionnaire.

3. What is confirmed by the study?

   A) Depression leads to excessive use of internet.

   B) Depression results from excessive use of Internet.

   C) Excessive use of Internet usually accompanies depression.

   D) Excessive use of Internet is usually prior to depression.

4. It is speculated by the public that online communities _____.

   A) can never replace normal social function

   B) are intended to replace normal social function

   C) are associated with psychological disorders

   D) shouldn't take the blame for psychological disorders

5. According to Dr. Catriona Morrison, the public speculation _____.

   A) lacks scientific evidence   B) helps clarify their study

   C) turns out to be correct   D) is worth further study

Part Four: Translation Skills

正译法和反译法

正译法和反译法通常用于汉译英，偶尔也用于英译汉。所谓正译，是指把句子按照与原语相同的语序或表达方式译入目的语。所谓反译则是指把句子按照与原语相反的语序或表达方式译入目的语。正译和反译常常具有同义的效果，但反译往往符合英语的思维方式和表达习惯，因此比较地道。

例1： 在美国，人人都能买到枪。

正译： In the United States, everyone can buy a gun.

反译： In the United States, guns are available to everyone.

详解： 反译体现出英语更倾向于用被动句表示句子。人买枪，符合汉语习惯，但枪能被大多数人所获得，则符合英语逻辑。available 就表示可获得的意思。

例2： 他突然想到了一个新主意。

正译： Suddenly he had a new idea.

He suddenly thought out a new idea.

反译： A new idea suddenly occurred to him.

详解： 这里体现了英语的惯用语序。他想到一个主意，符合汉语习惯，但一个主意突然迸发到人的脑海中，符合英语逻辑。A new idea suddenly occurred to him. 这个翻译更地道。sth. occurs to sb. 是英文的地道用法。

例3：你可以从因特网上获得这一信息。

正译：You can obtain this information on the Internet.

反译：This information is accessible/available on the Internet.

详解：这里体现了英语更倾向于用被动句表示句子的含义。你从因特网获得信息，符合汉语习惯，但信息在因特网上可获取或获得，则符合英语逻辑。available 或 accessible 表示可获得或是通过某种途径可以接触到的意思，词和中文一一对应，且符合英语说法，更显地道。

Translation Exercise 1

墙面装饰展现的是装饰工作的主要部分，极易给人留下第一视觉和第一印象，这对装饰质量的好坏至关重要的。装饰工程是为了美化建筑物、美化环境，带给人类无穷无尽的享受，因此，每个装饰工人都应该具有较强的责任感，做好自己的本职工作。

Translation Exercise 2

因工程调整而产生的工程洽商，如洽商费用超过五万元人民币，则洽商文件必须在工程中期验收当日被提交给甲方，所产生的洽商费用在交付中期时一并核算结清。如洽商费用低于五万元人民币，则洽商文件必须在工程验收当日内提交给甲方，所产生的费用在交付工程验收款时一并核算结清。

Unit 7 Tender Documents and Contracts

Learning Objectives

After completing this unit, you will be able to do the following:
- √ Grasp the main idea and the structure of the text;
- √ Master the key language points and grammatical structure in the text;
- √ Understand the basic knowledge of tender documents and contracts;
- √ Conduct a series of reading, listening, speaking and writing activities related to the theme of the unit.

Outline

The following are the main sections in this unit.
1. Warm-up Activity
2. Text
3. Words and Expressions
4. Exercises

Terms of Tender Documents and Contracts

In this unit, you will learn the meanings of terms listed below.
Quality and Safety Programs
Bill of Quantities
Commencement Date and Completion Date

Vocabulary

Listed below are some words appearing in this unit that you should make them a part of your vocabulary.
prequalification
representation
document
bid (bade, bidden)
authorize
attachment

Looking Ahead

The bidding process:

1. Preparation of tender document;
2. Preparation of prequalification document;
3. Prequalification of tenders;
4. Obtaining tenders;
5. Opening of tenders;
6. Evaluation of tenders;
7. Award of contract (performance security).

Introduction

As civil engineering works are often complex, involving the contractor in many hundreds of different operations using many different materials and manufactured items, including employment of a wide variety of specialists, the documents defining the contract are complex and comprehensive.

The task of preparing them for tendering therefore warrants close attention to detail and uniformity of approach, so as to achieve a coherent set of documents which forms an unambiguous and manageable contract. A typical set of documents prepared for tendering will include the following.

1. Instructions to tenderers.
2. General and particular conditions of contract.
3. The specification.
4. Bill of quantities or schedule of prices.
5. Tender and appendices.
6. The contract drawings.

Warm-up Activity

Do you know what tender documents and contracts are?

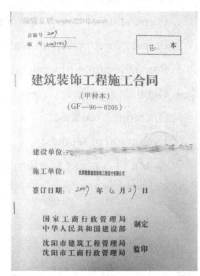

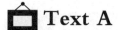

Text A

Tender Documents and Contracts

A contractor should pass the prequalification procedure first. As the document requires, in the first page he should write his company's full name, the legal representative, company's address, telephone and fax numbers, post number and e-mail address if it is available. The contractor fills in other forms as well. For example, for his employees' conditions, he should put the numbers of senior technicians, management staff and skilled workers. Regarding the construction equipments he should list the machines and other equipments owned by his company with their quantities, standards, depreciation years and producers' names. And if the contractor can provide a certificate from an honorable bank, saying that his company has a good credit record in the bank, it will be very helpful.

After the employer examines the prequalification documents, the company can be accepted to enter the next stage for tendering of the project.

The bidding document comprises four parts.

Part 1 Letter of Agreement.

Part 2 Construction Method Statement.

Part 3 Quality and Safety Programs.

Part 4 Bill of Quantities.

The Letter of Agreement is a promise of the contractor to the employer. It mentions the total initial contract price with the capital letters. The commencement date, the completion date and the maintenance period are also mentioned. This letter must be signed by the legal representative of the contractor or his authorized person.

Construction Method Statement consists of the following attachments: construction method description, construction schedules, site organization chart, key persons' information, workmen schedules, construction equipment schedules, layout for production area and living area, charts of electricity and water requirement, list of main supplier and subcontractors, list of repair parts, and so on.

The third part includes quality and safety programs. Certain measures are to be taken for safety assurance, and test reports are to be submitted. The contractor should prepare his own safety schemes according to the requirements.

The last part is the most important one. Each unit price should include material cost, labor cost, machine cost and the indirect costs, which include the items of staff salary, office fees, camp cost, transportation expenses for workers, charges from a higher level, profit, taxes, bank guarantees, insurance, and so on.

Words and Expressions

prequalification	[pri͵kwɔlɪfɪ'keɪʃən]	n. 资格预审
representation	[͵reprɪzen'teɪʃ(ə)n]	n. 代表，代理人
fax	[fæks]	n. 传真
regarding	[rɪ'ɡɑːdɪŋ]	prep. 关于
depreciation	[dɪ͵priːʃɪ'eɪʃ(ə)n]	n. 折旧
credit	['kredɪt]	n. 信用
document	['dɒkjʊm(ə)nt]	n. 文件，公文
bid (bade, bidden)	[bɪd]	v. 出价，投标
initial	[ɪ'nɪʃəl]	adj. 最初的，初期的
capital	['kæpɪt(ə)l]	n. 首都；资金；大写字母 adj. 主要的

commencement	[kə'mensmənt]	n. 开始
completion	[kəm'pli:ʃn]	n. 完结，竣工
authorize	['ɔθəraɪz]	vt. 授权，指定
attachment	[ə'tætʃmənt]	n. 附件
submit	[səb'mit]	vi. 提交
fill in		填写
as well		又，也，还

Exercises

Part One : Special Terms

1. Quality and Safety Programs　　_____
2. Bill of Quantities　　_____
3. Commencement Date and Completion Date　_____
4. 证书　　_____
5. 投标书　　_____
6. 协议书　　_____

Part Two: Situational Conversation

A: Good afternoon! Nice to meet you!

B: Good afternoon! Nice to meet you too! Welcome to our company for tendering of the power station project. Here is the tender document for the works.

A: There are so many volumes.

B: Yes. Volume 1 is the Contract Conditions, Volume 2 is the Bill of Quantities, Volume 3 is Scope of Works, Volume 4 and Volume 5 are Technical Specifications.

A: There must be a lot of requirements for technical issues, otherwise these two volumes wouldn't be so thick.

B: You are right. By the way, the engineers and supervisors working on this project should not only be able to read and use the technical specifications, but also remember that English is the contract working language. The technical and supervisory staff, even your foremen for the site, should be able to speak English. Now let's go on to the Tender Documents.

A: We noticed that Volume 6 is Quality Assurance. Would you give us some explanation about it?

B: For this project, there are 3 grades of Quality Assurance defined. The first grade applied in the structures is directly connected with the power generating such as turbine pedestals and boiler frame. The second grade is applied for the structures which play an important role in the power plant such as main power building and circular water pumping house. The third grade is the same as normal civil works structures. You are required to submit your Quality Assurance Program in accordance with these grades.

A: Yes, we will. Are there any more volumes?

B: Yes. The next is Volume 7, Drawings. Let's talk about it tomorrow, shall we?

Notes：

1. Volume 1 is the Contract Conditions, Volume 2 is the Bill of Quantities, Volume 3 is Scope of Works, Volume 4 and Volume 5 are Technical Specifications. 卷1是合同条件，卷2是工程量清单，卷3是工程范围，卷4和卷5是技术规格书。

2. For this project, there are 3 grades of Quality Assurance defined. 本项目定义了三种级别的质量保证等级。defined 是过去分词作定语。

3. The first grade applied in the structures is directly connected with the power generating such as turbine pedestals and boiler frame. 第一级用于与发电直接有关的结构，如汽轮机基座和锅炉炉架。applied in the structure 是过去分词短语作定语，在翻译时要尽量简练，符合汉语的习惯。

Exercise 1: Sentence Patterns

1. Welcome to our company for tendering of the power station project.

欢迎来本公司参加发电厂项目的投标。

2. Here is the tender document for the works.

这是本项目的投标书。

3. By the way, the engineers and supervisors working on this project should not only be able to read and use the technical specifications, but also remember that English is the contract working language.

顺便说一下，参与本项目工作的工程师和监理不仅要能阅读和运用这些技术条件，还要记住英语是本合同的工作语言。

4. Would you give us some explanation about it?

对此，你能向我们解释一下吗？

5. Let's talk about it tomorrow, shall we?

我们明天再谈好吗？

Exercise 2: Complete the following Dialogue in English

A: Nice to meet you!

B: Nice to meet you too! _____

_____.

（欢迎来本公司参加发电厂项目的投标。这是本项目的投标书。）

A: There are so many volumes.

B: _____
_____.

（是的。卷 1 是合同条件，卷 2 是工程量清单，卷 3 是工程范围，卷 4 和卷 5 是技术规格书。）

A: There must be a lot of requirements for technical issues, otherwise these two volumes wouldn't be so thick.

B: _____
_____.

（是这样。顺便说一下，参与本项目工作的工程师和监理不仅要能阅读和运用这些技术条件，还要记住英语是本合同的工作语言。）

The technical and supervisory staff, even your foremen for the site, should be able to speak English. Now let's go on to the Tender Documents.

A: We noticed that Volume 6 is Quality Assurance. _____

（对此，你能向我们解释一下吗？）

B: Yes, of course.

Part Three: Reading Comprehension

If you're like most people, you're way too smart for advertising. You skip right past newspaper ads, never click on ads online and leave the room during TV commercials.

That, at least, is what we tell ourselves. But what we tell ourselves is wrong. Advertising works, which is why, even in hard economic times, Madison Avenue is a $34 billion-a-year business. And if Martin Lindstrom — author of the best seller Buylogy and a marketing consultant for Fortune 500 companies, including Pepsi Co. and Disney — is correct, trying to tune this stuff out is about to get a whole lot harder.

Lindstrom is a practitioner of neuromarketing（神经营销学）research, in which consumers are exposed to ads while hooked up to machines that monitor brain activity, sweat responses and movements in facial muscles, all of which are markers of emotion. According to his studies, 83% of all forms of advertising principally engage only one of our senses: sight. Hearing, however, can be just as powerful, though advertisers have taken only limited advantage of it. Historically, ads have relied on slogans to catch our ear, largely ignoring everyday sounds — a baby laughing and other noises our bodies can't help paying attention to. Weave this stuff into an ad campaign, and we may be powerless to resist it.

To figure out what most appeals to our ear, Lindstrom wired up his volunteers, then played them recordings of dozens of familiar sounds, from McDonald's wide-spread "I'm Lovin' It" slogan to cigarettes being lit. The sound that blew the doors off all the rest—both in terms of interest and positive feelings—was a baby giggling. The other high-ranking sounds were less original but still powerful. The sound of a vibrating cell phone was Lindstrom's second-place finisher. Others that followed were an ATM dispensing cash and a soda being popped and poured.

In all of these cases, it didn't take an advertiser to invent the sounds, combine them with meaning and then play them over and over until the subjects being part of them. Rather, the sounds already had meaning and thus triggered a series of reactions: hunger, thirst, happy anticipation.

1. As is mentioned in the first paragraph, most people believe that _____.

    A) ads are a waste of time   B) ads are unavoidable in life

    C) they are easily misled by ads   D) they are not influenced by ads

2. What do we know about Madison Avenue in hard economic times?

    A) It becomes more thriving by advertising.

    B) It turns to advertising so as to survive.

    C) It helps spread the influence of advertising.

    D) It keeps being prosperous thanks to advertising.

3. What do we learn about Pepsi Co. and Disney from the passage?

    A) Lindstrom was inspired by them to write a book.

    B) They get marking advice from Lindstrom.

    C) Lindstrom helps them to go through hard times.

    D) They attribute their success to Lindstrom.

4. It is pointed out by Lindstrom that advertisers should _____.

    A) rely more on everyday sounds

    B) rely more on neuromarketing

    C) rely less on slogans

    D) rely less on sight effect

5. It is found by Lindstrom that a baby giggling is _____.

    A) the most touching   B) the most familiar

    C) the most impressive   D) the most distinctive

Part Four: Translation Skills

倒置法

在汉语中，定语修饰语和状语修饰语往往位于被修饰语之前，而在英语中，许多修饰语则常常位于被修饰语之后，因此翻译时往往要把原文的语序颠倒过来。倒置法通常用于英译汉，即对英语长句按照汉语的习惯表达法进行前后调换，按意群或进行全部倒置，原则是使汉语译句符合现代汉语论理叙事的一般逻辑顺序。如：

例1： At this moment, through the wonder of telecommunications, more people are seeing and hearing what we say than on any other occasions in the whole history of the world.

翻译：此时此刻，通过现代通信手段的奇妙，能看到和听到我们讲话的人比史上这世界中任何其他场合的人都要多。（部分倒置）

详解：根据意义上的需要，可以在主语和谓语之间进行倒装。如果把主语 more people 直接放在句首，例1中的 seeing and hearing what we say 的翻译似乎不够明确。通过主语和谓语的部分倒装，把谓语动词提到前面来，意思更为明确和清晰，读起来也较通顺自然，符合汉语习惯。

例2： I believe strongly that it is in the interest of my countrymen that Britain should remain an active and energetic member of the European Community.

翻译：我坚信，英国依然应该是欧共体中的一个积极的和充满活力的成员，这是符合我国人民利益的。（部分倒置）

详解：根据意义上的需要，可以把两个从句进行倒装。如果把第一个从句 that it is in the interest of my countrymen 的翻译直接放在句首，则 that Britain should remain an active and energetic member of the European Community 的翻译似乎不够明确。通过把两个从句进行倒装，意思更为明确和清晰，读起来也较通顺自然。

例3： Great changes have taken place in China since the introduction of the reform and opening policy.（全部倒置）

翻译：改革开放以来，中国发生了巨大的变化。

详解：根据意义上的需要，可以在主语和谓语之间进行倒装。如果把主语 Great changes 直接放在句首翻译，则 have taken place in China 的翻译似乎不够明确。通过主语和谓语的部分倒装，把谓语动词提到前面来，意思更为明确和清晰，读起来也较通顺自然，符合汉语习惯。

Translation Exercise 1

协议书是承包商对业主的承诺。协议书应用大写方式写清初始合同的总价格，以及开工日期、竣工日期和保修期。这份协议书必须由承包商的法人代表或其授权人签名。

Translation Exercise 2

施工方案中包含以下附件：施工方案内容、施工进度、现场布置图、主要人员资质、工人们的安排、施工设备清单、生产区域和生活区域布局、电力和水需求布置、主要供应商和分包商的名单、备件清单等。

Unit 8 Construction

Learning Objectives

After completing this unit, you will be able to do the following:
- √ Grasp the main idea and the structure of the text;
- √ Master the key language points and grammatical structure in the text;
- √ Understand the basic knowledge of construction;
- √ Conduct a series of reading, listening, speaking and writing activities related to the theme of the unit.

Outline

The following are the main sections in this unit.
1. Warm-up Activity
2. Text
3. Words and Expressions
4. Exercises

Terms of Construction

In this unit, you will learn the meanings of terms listed below.
working efficiency
warehouse
crusher

Vocabulary

Listed below are some words appearing in this unit that you should make them a part of your vocabulary.
perform
evaluate
equipment
mixer
automate

Looking Ahead

Construction

Construction is the process of constructing a building or infrastructure. Construction differs from manufacturing in that manufacturing typically involves mass production of similar items without a designated purchaser, while construction typically takes place on location for a known client. Construction as an industry comprises six to nine percent of the gross domestic product of developed countries. Construction starts with planning, design, and financing; and continues until the project is built and ready for use.

Introduction

The construction work can be divided into a number of stages:

1. Evaluation of plans, specifications, basic demands and features of the site;
2. Plan and speed of the job;
3. Making the site ready;
4. Building the structure;
5. Cleaning up.

The first stage of evaluation consists of a careful study of demands of design and of the site itself. Too often this is not done until the third and fourth stages are under way, which is far too late.

The second stage is most important if the job is to be done economically. The equipments, labor and materials for each stage in the construction must be provided at the correct time.

The third stage includes constructing access roads, making warehouse, crushers, concrete mixers, offices and housing for the workmen ready. Of course, this work is often just the beginning, the arrangements are changed several times during the progress of the work. The major part of the time and money is spent on the building stage.

Warm-up Activity

Do you know what construction is?

Text

Construction

Construction is the translation of a design to reality. It is as important and complicated as the design in order that the structure will perform as it was intended to and the work is finished within the required time at the lowest cost.

The designer must be in close contact with everything that is done during the construction work so that any changes in the site conditions, materials and work being done can be evaluated and, if necessary, corrected or improved.

The constructors should have the same knowledge of the working-plan as the designer. They must also know the details of the design and must understand any unusual aspect of the design. In fact, the constructors should go to the engineers for information and advice during the design stage so that the plans do not call for something that cannot be built economically. Both the designers and constructors must always work in harmony.

More workers are employed during the peak period. The employees should be given training for working skills and knowledge about quality and safety as early as possible. It will improve the working efficiency a lot.

The construction work can be divided into a number of stages:

1. Evaluation of plans, specifications, basic demands and features of the site;
2. Plan and speed of the job;
3. Making the site ready;

4. Building the structure;

5. Cleaning up.

The first stage of evaluation consists of a careful study of demands of design and of the site itself. Too often this is not done until the third and fourth stages are under way, which is far too late.

The second stage is most important if the job is to be done economically. The equipment, labor and materials for each stage in the construction must be provided at the correct time.

The third includes constructing access roads, making warehouse, crushers, concrete mixers, offices and housing for the workmen ready. Of course, this work is often just the beginning, the arrangements are changed several times during the progress of the work. The major part of the time and money is spent on the building stage.

With the development of science and technology, construction methods will change in all areas. Many daily functions will become automated and computer-controlled, especially in residential construction. Thus there will be a demand for more skilled workmen, primarily those having technical background.

Words and Expressions

reality	[rɪˈælɪtɪ]	n. 现实
perform	[pəˈfɔːm]	v. 执行，完成（事业）
intend	[ɪnˈtend]	vt. 打算
require	[rɪˈkwaɪə]	vt. 需要，要求
requirement	[rɪˈkwaɪəm(ə)nt]	n. 需要，要求
condition	[kənˈdɪʃ(ə)n]	n. 条件
evaluate	[ɪˈvæljʊeɪt]	vt. 估……的价
harmony	[ˈhɑːmənɪ]	n. 协调，和谐
employ	[ɪmˈplɒɪ]	vt. 雇用，聘用
peak	[piːk]	n. 山峰，高峰
employee	[ɪmplɔɪiː]	n. 雇员，员工
evaluation	[ɪˌvæljuˈeɪʃn]	n. 评价
equipment	[ɪˈkwɪpm(ə)nt]	n. 装备，设备，器材
crusher	[ˈkrʌʃə]	n. 轧碎机，碎石机

mixer	['mɪksə]	n.	搅拌机
arrangement	[ə'reɪn(d)ʒm(ə)nt]	n.	安排
automate	['ɔ:təmeɪt]	vt.	使自动化
residential	[rezɪ'denʃ(ə)l]	adj.	住宅的

in order that 目的在于
within the required time 在规定的时间内
at the lowest cost 以最低的成本
be in contact with 跟……保持联系
the same ... as 同……一样
in fact 实际上
call for 要求，需求
in harmony 和睦，融洽
divide into 分为，分成
be under way 在进行
spend...on 把……花在……上

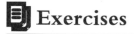 Exercises

Part One : Special Terms

1. working efficiency　_____
2. warehouse　_____
3. crusher　_____
4. 在要求的时间内　_____
5. 以最低的成本　_____
6. 施工人员　_____

Part Two : Situational Conversation

A: How many people is the project management composed of ?

B: About 200. The top management has three persons only. They are Project Manager, Deputy Project Manager and Chief-Engineer.

A: How many departments are there under the project management?

B: Five. They are Works, Technology, Goods, Administration and Quality Departments.

A: What about the function of each department?

B: The Works Department is responsible for all construction sections and terms. The Deputy Project Manager is also the manager of Works Department. The Technology Department is in charge of all matters about drawing and designing of temporary facilities. The Goods Department is in charge of material, construction equipment supply and the warehouse. The Administration Department is in charge of all matters related to finance and public relationship. The Quality Department is a special agency for it is controlled by not only Project Management but also the Headquarters of the company.

A: Can you tell us something about the key persons for the Project Management?

B: Sure. They are all qualified engineers with at least 10-year experience in the construction field. Here are their resumes including the proposed positions for the project, their education degrees, previous experience, especially the projects in which they have participated.

A: I want to remind you that as long as these key staff members are approved to work for the project, they should not be removed from the site without our permission. By the way, I also want to know something about your foremen.

Notes:

1. The Administration Department is in charge of all matters related to finance and public relationship. 行政部门负责财务和公关事务。related to … 是过去分词短语作后置定语。

2. …as long as these key staff members are approved to work for the project, they should not be removed from the site without our permission. ……只要这些主要人员被批准为此工程工作后，不经我们的允许，他们不能擅自调离。

Exercise 1: Sentence Patterns

1. What about the function of each department?
每个部门的职责是怎样的？

2. The Works Department is responsible for all construction sections and terms.
工程部门负责所有的施工步骤和条款。

3. The Technology Department is in charge of all matters about drawing and designing of temporary facilities.
技术部门掌管临时设施的绘制和设计所有事务。

4. Can you tell us something about the key persons for the Project Management?
你能给我们说说项目管理的重要成员吗？

5. By the way, I also want to know something about your foremen.

顺便说一下，我还想了解一下你们的工头。

Exercise 2: Complete the Following Dialogue in English

A: _____?

（该项目管理层由多少人组成？）

B: About 200. The top management has three persons only. They are Project Manager, Deputy Project Manager and Chief-Engineer.

A: _____?

（该项目管理下有多少个部门？）

B: Five. They are Works, Technology, Goods, Administration and Quality Departments.

A: _____?

（每个部门的职责是怎样的？）

B: The Works Department is responsible for all construction sections and terms. The Deputy Project Manager is also the manager of Works Department. The Technology Department is in charge of all matters about drawing and designing of temporary facilities. The Goods Department is in charge of material, construction equipment supply and the warehouse. The Administration Department is in charge of all matters related to finance and public relationship. The Quality Department is a special agency for it is controlled by not only Project Management but also the Headquarters of the company.

A: _____?

（你能跟我们说说项目管理的重要成员吗？）

B: Sure. They are all qualified engineers with at least 10-year experience in the construction field. Here are their resumes including the proposed positions for the project, their education degrees, previous experience, especially the projects in which they have participated.

A: I want to remind you that as long as these key staff members are approved to work for the project, they should not be removed from the site without our permission. _____?

（顺便说一下，我还想了解一下你们的工头。）

Part Three: Reading Comprehension

No document is safe any more. Faking once the domain of skilled deceivers that used expensive engraving（雕刻）and printing equipment, has gone mainstream since the price of desktop-publishing systems has dropped. In ancient times, faking was a hanging offence. Today, desktop counterfeiters have little reason to worry about prison, because the systems they use are universal and there is no means of tracing forged documents to the machine that produced them. This, however, may soon change thanks to technology development by George Chiu, an anti-faking engineer.

His approach is based on detecting imperfections in the print quality of documents. Old-school court scientist were able to trace documents to particular typewriters based on quirks（瑕疵）of the individual keys. He employs a similar approach, exploiting the fact that the rotating drums and mirrors inside a printer are imperfect pieces of engineering which leave unique patterns of banding in their products.

Although these patterns are invisible to the naked eye, they can be detected and analyzed by computer programs, and it is these patterns that Dr. Chiu has spent the past year devising. So far, he cannot trace individual printers, but he can tell pretty reliably which make and model of printer was used to create a document.

That, however, is only the beginning. While it remains to be seen whether it will be possible to trace a counterfeit（伪造）document back to its guilty creator on the basis of manufacturing imperfections, Dr. Chiu is now working out ways to make those imperfections deliberate. He wants to modify the printing process so that unique, invisible signatures can be incorporated into each machine produced which would make any document traceable.

Ironically, it was after years of collaborating with printing companies to reduce banding and thus increase the quality of prints, that he came up with the idea of introducing artificial banding that could encode identification information into a document. Using the banding patterns of printers to secure documents would be both cheap to implement and hard, if not impossible, for those without specialist knowledge and hardware to hide out.

Not surprising, the American Secret Service is monitoring the progress of this research very closely, and is providing guidelines to help Dr. Chiu to travel in what the service thinks is the right direction, which is fine for catching criminals. But how the legal users of printers will react to Big Brother being able to track any document back to its source remains to be seen.

1. By saying "no document is safe any more", the author probably means _____.

    A) cheap printers make it possible for anyone to forge documents

    B) the American Secret Service will be able to trace any document

C) every printed document will be secretly marked out through high-tech

D) counterfeiters have more advanced technology to use

2. The core of both old and new ways of anti-counterfeiting is _____.

A) the quirks of the keys of the typewriters

B) the drums and mirrors of the printers

C) the subtle flaws of printing devices

D) the special skills of the experts

3. According to Dr. Chiu, what makes it possible to track down any paper work is to _____.

A) make all the imperfections of machines deliberately

B) find out the guilty creator of the counterfeit document

C) allow more imperfectly designed printers to be sold

D) mark all the printers in a special and secret way

4. The advantage of using banding patterns to trace documents is that it _____.

A) is economically affordable and technically practical

B) strengthens the collaboration among printing companies

C) makes printers cheaper and hard to be taken around

D) is able to identify even the most specialized criminals

5. It can be inferred from the last paragraph that _____.

A) Dr. Chiu will be remembered for his special contribution

B) the tracking of all documents might result in controversies

C) American Secret Service is funding Dr. Chiu's research

D) printer manufacturers are reluctant to implement deliberate banding

Part Four: Translation Skills

包孕法

这种方法多用于英译汉。所谓包孕是指在把英语长句译成汉语时，把英语后置成分按照汉语的正常语序放在中心词之前，使修饰成分在汉语句中形成前置包孕。但修饰成分不宜过长，否则会造成拖沓或汉语句子成分在连接上的纠葛。如：

例1：You are the representative of a country and of a continent to which China feels particularly close.

翻译：您是一位来自于使中国倍感亲切的国家和大洲的代表。

详解：进行翻译有时需要全面考虑从句的意思，比如例1直译就是"您是一位来自于使中国倍感亲切的国家的代表"，意思并不完整。

例2：What brings us together is that we have common interests which transcend those differences.

翻译：使我们走到一起的，是我们有超越这些分歧的共同利益。

详解：进行翻译有时需要全面考虑从句的意思，比如该例如果直译就是"使我们走到一起的，是共同利益"，那么这个定语"是我们有超越这些分歧的"就没有翻译出来，意思并不完整。

Translation Exercise 1

施工是把设计转化为实体。它与设计一样重要和复杂，以便设计内容能如预期一样变为现实，并且该工作要在规定时间内以最低的成本完成。设计师必须密切接触施工工作中的一切事务，以便对现场条件、材料以及进行中的作业出现的任何更改进行评估，如有必要，还要进行修正或改进。

Translation Exercise 2

施工人员应当和设计师一样具有施工平面图方面的知识。他们还必须知道设计的细节，必须了解任何例外的设计情况。事实上，施工人员在设计阶段应该向工程师寻求信息和建议，以便设计上不致出现某些在经济上不能承受的项目。设计师和施工人员必须协调一致地开展工作。

练习答案及参考译文

Unit 1　Civil Engineering

Part One: Special Terms

1. 区别/区分 A 和 B

2. 对……产生影响

3. 满足……的需求

4. construction management

5. basic research/investigation

6. municipal works/engineering

Part Two: Situational Conversation

Exercise 2: Complete the Following Dialogue in English

A: an America visiting scholar

B: a Chinese student majoring in engineering cost

A: How do you do ?

B: How do you do ?（您好。）

A: May I ask you some questions?

B: Of course/ Certainly.（当然可以。）

A：What do you do?

B: I am a student in this university.（我是这个学校的一名学生。）

A: Can you tell me what major you are studying?

B: Engineering cost.（工程造价专业。）

A: How do you think of this major?

B: It is a hot/promising major these years, and I like it.（前景广阔，我喜欢。）

A: How many courses are there in this major?

B: There are more than ten courses, which are compulsory/required and basic/fundamental courses.（十多门课程，主要是专业必修课和基础课。）

A: Which course do you like best? And why ?

B: Construction Techniques. Because the teacher can clarify the lesson from the shallower to the deeper and from the easier to the more advanced, and integrate theory with practice.（建筑施工技术，因为老师讲课由浅入深，由易入难，并能理论联系实际。）

A: Thank you very much.

B: <u>You are welcome.</u>（不客气。）

Part Three: Reading Comprehension

1-5 BDCDC

Part Four: Translation Skills

Translation Exercise 1

In the university, mathematics is heavily emphasized throughout the engineering curriculum. Today, it includes courses in statistics, which deals with gathering, classifying and using numerical data or pieces of information. An important aspect of statistical mathematics is probability, which deals with what may happen when there are different factors or variables that can change the results of a problem. Before the construction of a bridge is undertaken, for example, a statistical study is made of the amount of traffic the bridge will be expected to handle. In the design of the bridge, variables such as water pressure on the foundation, impact, the effects of different wind forces and many other factors must be considered.

Translation Exercise 2

There are four main types of work in building production. They are civil construction workers, building installers, building mechanics and building decorators. Of course, each of these includes more than eight types. For example, building installers can be divided into plumber, electrician, welder, hoister, ventilating workers, riveter, fitter, and so on. The tradesman of each type needs to work in close cooperation, not a single one of them can be dispensed with. Each tradesman plays an important part in construction.

参考译文

土木工程

土木工程是指建筑环境的规划、设计、施工和管理等一系列活动。这个环境包括从灌溉和排水系统到火箭发射设施的所有根据科学原理建造的结构物。

土木工程师修建道路、桥梁、隧道、大坝、港口、发电站、水系统和污水系统、医院、学校、公共交通系统，以及现代化社会和大量人口集中的地方所必需的其他公共设施。他们也修建私人设施，如机场、铁路、管线、高楼大厦和为工业、商业、民用设计的其他大型建筑。此外，土木工程师规划、设计和修建整个城市和乡镇，最近已经开始规划和设计空间站以容纳独立的（科研）团体。

"土木"这个词是从拉丁语"citizen"派生而来的。1782年，英国人 John Smeaton 用这个术语来区分他的非军事工程项目和当时占统治地位的军事工程师的工程项目。从那以后，土木工程这个术语便被用来指那些修建公共设施的工程师，尽管这个领域涉猎的范围比以前更广。

范围

因为土木工程的范围太广，所以它被细分为许多技术专业。根据工程的类型，土木工程师需要各种技能。当一项工程开始时，土木工程师要勘测现场并绘图，他们还要确定该项目所在土层的承载力。岩土工程专家要做土工试验以确定该土地是否能承受这项工程的重量。环境专家要研究工程对当地区域的影响：潜在的空气污染和地下水污染，工程对当地动植物的影响，以及工程怎样设计才能满足政府对保护环境的要求。运输专家要确定需用什么类型的设施来减轻由完工的工程产生的荷载对当地道路和其他运输网带来的压力。同时，结构专家用初始资料来做工程的详细设计、规划和说明书。从工程开始到结束，施工管理专家监督和协调这些土木工程专家们的工作。根据其他专家提供的信息，施工管理土木工程师要估计材料和劳动力的数量和成本，安排所有的工作，订购工作所需的材料和设备，雇承包商和分包商，以及做其他的监督管理工作以确保工程能够按照说明按时完工。

对于任何给定的工程，土木工程师都广泛地使用计算机。计算机被用来设计工程的各个部分（即计算机辅助设计，简称 CAD）并进行管理。计算机对于现代土木工程师而言是必不可少的，因为它们可使工程师高效地处理大量数据，这些数据是在确定最优施工方案时所需要的。

结构工程　在这个专业里，土木工程师规划和设计所有类型的结构，包括桥梁、大坝、电站、设备的支撑、近海工程的特殊结构、美国的太空计划、发射塔、巨型天文望远镜和无线电望远镜以及许多其他工程。结构工程师用计算机确定结构必须抵抗的力：自重、风力、引起建筑材料膨胀或收缩的温度变化以及地震程度。他们还确定适当的材料组合：钢材、混凝土、塑料、石料、沥青、砖、铝或其他的建筑材料。

水资源工程　该专业的土木工程师处理水的自然调节的各个方面。该工程有助于阻挡洪水，为城市和灌溉系统供水，管理和控制河流流量，维修河滩和其他滨水区的设施。此外，他们还设计和维修港口、运河和船闸，修建大型水力发电大坝和小型水坝以及各种类型的围堰。他们还帮助设计海上建筑物，以及确定影响航运结构物的位置。

岩土工程　专攻此领域的土木工程师分析支撑结构物并影响结构性能的土壤和岩石特性。他们计算建筑和其他结构由于自重压力可能引起的沉降，并采取措施使之减少到最小。这些工程师还估算并确定怎样加强斜坡和填方以及怎样保护结构免遭地震和地下水的影响。

管道工程　在土木工程的这个分支里，工程师修建运输液体、气体或固体的管道和相关的设施，运输的物质范围从煤浆和半液体废料到水、石油和不同类型的高燃性和非燃性气体。工程师要确定管道的设计，工程对所穿过地区的经济和环境影响，要用到的材料类型——钢材、混凝土或不同材料的组合的安装技术，检测管道强度的方法，怎样控制以保持适当的压力，以及正在被运送材料的流通速度。当输送危险材料时，安全也是应该考虑的主要问题。

  环境工程 在工程的这个分支里，土木工程师设计、修建和监督各个系统以提供安全的饮用水，防止和控制地表水和地下水供应的污染。他们也设计、修建和监督各项工程以控制或消除土地和空气污染。这些工程师修建水厂和污水处理厂，设计空气净化器和其他设备以减少或消除由工业加工、焚烧或其他一些产生烟雾的行为导致的空气污染。他们也采取措施，通过修建专门的垃圾场或者进行有毒有害物质的无害化处理来控制有毒有害废弃物。此外，工程师们对垃圾填埋进行设计和管理以防止其对周围土地的污染。

  运输工程 从事这一专业领域的土木工程师建造一些设施以确保人和货物的安全和高效的运输。他们专门研究设计和维护各种类型的运输设施，公路和街道，公共交通系统，铁路和机场，港口及海港。运输工程师在设计每一个项目的过程中，既要运用技术知识，也要考虑经济、政治和社会的因素。他们与城市规划者紧密配合，因为社区的质量直接关系到运输体系的质量。

  建筑工程 此领域的土木工程师从头至尾一直监督工程的施工。他们有时被称为项目工程师，他们不仅运用技术技能，还运用管理技能，包括施工方法、规划、组织、筹集资金和项目管理施工方面的知识。事实上，他们协调工程中每个人的活动，如勘测员，为临时道路和斜坡定线和施工、挖基础、建模和浇注混凝土的工人以及绑扎钢筋的工人。这些工程师还为建筑业主定期提供进度报告。

  社区和城市规划 从事土木工程这一方面的工程师可能规划和发展一个城市中的社区或整个城市。此规划中所包括的远不仅是工程因素，土地的开发使用和自然资源环境、社会和经济因素也是主要的成分。这些土木工程师对公共建设工程的规划和私人建筑的发展进行协调。他们评估所需的设施，包括街道、公路、公共运输系统、机场、港口、给排水和污水处理系统、公共建筑、公园和娱乐及其他设施以保证社会、经济和环境的协调发展。

  摄影测量、勘测和制图 在这一专业领域的土木工程师精确测量地球表面以获得可靠的信息来定位和设计工程项目。这一方面包括高工艺学方法，如卫星成像、航拍和计算机成像。来自人造卫星的无线电信号，通过激光和音波柱扫描被转换为地图，为隧道钻孔、建造高速公路和大坝、绘制洪水控制和灌溉方案、定位可能影响建筑项目的地下岩石构成以及许多其他建筑用途提供更精准的测量。

其他的专门项目

还有两个并不完全在土木工程范围里面但对训练相当重要的附加专门项目是工程管理和工程教学。

工程管理 许多土木工程师都选择最终走向管理岗位的职业生涯。其他则能让他们的职业生涯从管理岗位开始。土木工程管理者结合技术上的知识和组织能力来协调劳动力、材料、机械和钱。这些工程师可能是工作在政府——市政、郡、州或联邦的人员；工作在美国陆军军团作为军队或民用工程的管理工程师；或在半自治地区机构，城市主管当局或相似的组织。他们也可能是管理规模从几个到上百个雇员的私营工程公司。

工程教学 通常选择教学事业的土木工程师教授研究生和本科生技术上的专门项目。许多从事教学的土木工程师参与到产生建筑材料和施工方法技术革新的基础研究。多数人也担任工程项目或技术领域的顾问和主要项目的代理。

Unit 2　Design Process

Part One: Special Terms

1. 定性分析

2. 骨架

3. 金属雕刻

4. remarkable similarities

5. quantitative analysis

6. metal framework

Part Two: Situational Conversation

Exercise 2: Complete the Following Dialogue in English

A: Excuse me! Can you answer me some questions about building drawings?

B: Certainly. What do you want to know about building drawings?

（当然可以。关于建筑图纸你想了解什么？）

A: First, I want to know the scientific definition of building drawings.

B: The so-called building drawings are the designs drawn minutely and accurately by the method of orthogonal projection in accordance with the relevant regulations for the proposed buildings in the content of the internal and external shape and size, the structure, construction, decoration and equipment of different parts, and so on.

（所谓的建筑图纸就是将一幢拟建建筑物的内外形状和大小，以及各部分的结构、构造、装饰、设备等内容，按照有关规范规定，用正投影方法详细准确地画出的图样。）

A: So building drawings are very important to the construction works, aren't they?

B: You are right. Building drawings have always been regarded as the norm and the guide of our construction works.

（你说的没错。建筑图纸一直被看作施工的指南和准则。）

A: How many parts are there in a complete set of building drawings?

B: According to the professional content and functions, a complete set of building drawings generally include the catalogue, the general design specification, the building construction drawing, the structure and construction.

（一套完整的施工图，根据专业内容或者作用不同，一般包括图纸目录、设计总说明、建筑施工图、结构施工图以及设备施工图。）

A: How can I tell the difference between the building construction drawings and the structural working drawings?

B: The building construction drawing specifies the inside layout, the external shape and the requirements of decoration, structural working drawing and equipment, etc.

（建筑施工图表示的是建筑物的内部布置情况、外部形状以及装修、构造、施工要求等。）

A: What about structural working drawings?

B: The structural working drawing specifies the layout, the type, the size and the construction of the load-bearing parts and a building.

（结构施工图表示的是承重结构的布置情况、构件类型、尺寸大小及结构做法等内容。）

A: To make these complicated drawings, some precise instruments are inevitable, am I right?

B: Yes. The commonly used instruments are the drawing pen, the drawing-compasses, the T-square, the French curve.

（是的，常用的绘图仪器有绘图笔、绘图圆规、丁字尺、曲线板等。）

A: I know. There are mainly used in manual drawings, aren't they?

B: Right. At present, the architectural drawings are mainly finished by the computer AutoCAD software, the advantages of which are self-evident in contrast to the manual drawing.（没错，现在的建筑图纸主要是通过 AutoCAD 软件绘制出来的。相对于手工绘图，其优点是不言而喻的。）

A: What is AutoCAD, I want to know?

B: AutoCAD is an automated computer aid design software developed by the America Autodesk Corporation for two-dimensional and three-dimensioned design and graphing.

（AutoCAD是美国Autodesk公司开发的用于二维及三维设计、绘图的自动计算机辅助软件。）

A: Will you mind teaching me to make drawings by AutoCAD something when you are free?

B: <u>Of course not. If necessary, you can give me a call, and I will do my best to help you</u>.（当然不介意，如有需要，尽管给我打电话，我一定会不遗余力地帮你。）

Part Three: Reading Comprehesion

1-5 CADCD

Part Four: Translation Skills

Translation Exercise 1

Building drawings involve a variety of drawings such as plan, elevation, cross section, layout, bird's eye view, standard drawing, working drawing... Drawings play an important part in construction. Therefore, they are regarded as the norm and the guide of the construction. In other word, construction is the translation of design into reality.

Translation Exercise 2

For the contractor, a bid estimate submitted to the owner either for competitive bidding or for negotiation consists of direct construction cost including field supervision, plus a markup to cover general overhead and profits. The direct cost of construction for bid estimates is usually derived from a combination of the following approaches: subcontractor quotations, quantity take-offs and construction procedures.

参考译文

设计过程简介

大多数优秀的景观建筑师都要经历反复分析和创造性思考过程，也就是设计过程，才能尽可能地以既经济又给人以美感的方式，使所完成的场地设计方案恰到好处地满足一切必需的要求。设计过程的作用不少，其中包括：

1. 为设计方案提供合理和规范的框架；
2. 确保形成的方案能适合给定的设计环境（场地、客户的要求、预算等）；
3. 有利于客户通过研究各种可行方案来决定场地的最佳用途；
4. 作为向客户阐明并维护其方案的依据。

设计过程有时也可以称为"解决问题的过程"。它包括若干步骤，这些步骤通常（并非必然）有先后顺序。一般说来，建筑师、工业设计师、工程师以及科学家们都采用同样的步骤来解决问题。对场地设计人员来说，其设计过程一般包括下列步骤。

1. 与客户（甲方、业主）签合同

2. 研究与分析（包括踏勘场地）

（1）基地平面图绘制。

（2）场地清单（数据收集）和分析（评价）。

（3）会见客户。

（4）制订计划。

3. 设计

（1）理想的功能分析图。

（2）场地相关功能分析图。

（3）概念设计。

（4）构图研究。

（5）初步设计。

（6）总体设计。

（7）详细设计。

4. 实施

（1）施工。

（2）绿化。

5. 养护

6. 完工后的评价

以上这些步骤是设计工作的理想过程。实际上，许多步骤都是相互重叠和兼容的，彼此的界线并不十分明显。况且，有些步骤彼此并列同时进行。例如，会见客户与制订计划可以同时进行，而场地踏勘与场地分析也可以同时进行；在另一些场合，可能有必要重复前一步骤。因为某个数据当初被忽视，或某个人对场地的印象需要加深，设计阶段开始之后，也会有必要重新踏勘场地或再次与客户洽谈。

设计新手必须懂得，完美而实用的设计方案不会像变魔术一样从天上掉下来。没有任何奥妙的公式或神秘的心境可以毫不费力地提出好的设计方案。而设计方案绝不是只用铅笔在纸上画画就可得出来的。

实用且给人以激情的设计要求做大量敏锐的观察、分析、研究、推敲和再推敲，外加一定的灵感和独创性。

应该注意，提出一个设计方案确实包括理性方面（调查、分析、计划编制和施工知识）和直觉方面（把式样和造型结合起来的感觉、美学欣赏能力等）。因此，设计的过程就是一系列设计步骤的总框架。它把理性和直觉两个方面结合起来，有助于设计人员组织自己的工作、思维和情感，尽可能作出最佳的设计方案。

Unit 3　Structural Materials

Part one: Special Terms

1. 抗拉强度

2. 抗压强度

3. 加强和预应力混凝土

4. plain (unreinforced) concrete

5. prestressed concrete

6. Portland cement

Part Two: Situational Conversation

Exercise 2: Complete the Following Dialogue in English

(A: a salesman　B: a customer)

A: Can I help you?

B: <u>Yes, I'd like to order some marbles.</u>（嗯，我打算订购一批石材。）

A: For what purpose?

B: <u>Mainly for housing decoration.</u>（家里装修用的。）

A: All the samples of stone materials are here. What do you want to buy?

B: <u>I am very interested in the marble here. If the price is reasonable, I can place the order with you right away.</u>（我对你这里的大理石很满意，如果价格合理的话，我现在就订货。）

A: I'm very glad to hear that.

B: <u>What is the lowest price for the pure white marble?</u>（这种纯白的大理石最低价是多少？）

A: 700 yuan per square meter.

B: I think the price is too high. Can you reduce it?

A: <u>I am afraid not. 700 yuan per square meter is our bottom price. Moreover, these are purely natural marbles, and each has its unique pattern and color.</u>（这恐怕不行，700元是我们的底价了。更何况这种大理石是纯天然的，每一块都有独一无二的图案和色彩。）

B: But there are some beautiful marbles in the market priced at 120-180 yuan per square meter. How do you explain the large price differences to me?

A: <u>Those at low prices are artificial marbels, with no good transparency or becoming lusterless.</u>（那些价格低的是人造大理石，透明度不好，且没有光泽。）

B: Oh, I got it. Can you tell me how to distinguish the two different kinds of marbles?

A: <u>The easiest way is simply to dip a few drops of diluted hydrochloric acid on the marbles, the natural ones will be effervescing vigorously, which the artificial ones will be effervescing weakly or will not be effervescing at all.（最简单的方法就是滴上几滴稀盐酸，天然大理石会剧烈起泡，人造大理石起泡弱甚至不起泡。）

B: Can I do a small experiment on the samples of marbles here by using diluted hydrochloric acid?

A: Of course, you can. As the saying goes, pure gold does not fear furnace.（当然可以。真金不怕火炼。）

(After the experiment)

B: Well, I'll accept the price and place an initial order of 10,000 square meters.

A: Very good! It's a pleasure to do business with you.（太好了。跟你做生意真是我的荣幸。）

B: The pleasure is ours. Can you deliver the goods by June 2nd?

A: No problem at all. We will not delay your business.（绝对没有问题，我们不会误事的。）

Part Three: Read Comprehension

1-5 DBDCB

Part Four: Translation Skills

Translation Exercise 1

A Joint Committee（联合委员会）on Concrete and Reinforced Concrete was established in 1904 by the American Society of Civil Engineers（土木工程师协会）, the American Society for Testing and Materials, the American Railway Engineering Association, and the Association of American Portland Cement Manufactures. This group was later joined by（加入）the American Concrete Institute. Between 1904 and 1910 the Joint Committee carried out research. A preliminary（初步的）report issued in 1913 lists the more important papers and books on reinforced concrete published（发表）between 1898 and 1911. The final report of this committee was published in 1916. The history of reinforced concrete building codes in the United States was reviewed（回顾）in 1954 by Kerekes and Reid.

Translation Exercise 2

Roof group fight: Roof patch shipped to the scene assembly; and assembly platform formation. Assembling and bagging chief roof should guarantee size requirements. Welding one side passed the inspection, then stand up and welded the other side, do construction records. Lifting the rear prospective admission experience. Roof and rain can also be a lifting assembled on the ground, but temporary reinforcement, hoisting to ensure we have the necessary stiffness.

参考译文

合适的结构材料的实用性是限制富有经验的结构工程师成就的主要原因之一。早期的建筑者几乎都只使用木材、石头、砖块和混凝土。尽管铸铁在埃及人修建金字塔时已被使用，但是把它作为建筑材料却由于大量熔炼的困难度而被限制。借由产业革命，然而，随着工业革命的到来，铁作为建筑材料和冶炼材料的双重需求变得大量起来。

英国土木工程师 John Smeaton 在 18 世纪中叶率先广泛地使用铸铁作为建筑结构材料。在 1841 之后，可锻金属发展成为更可靠的材料并且得到广泛的应用。尽管可锻金属优于铸铁，但仍有很多结构破坏从而需要更可靠的材料。钢便是这一需要的答案。1856 年的贝色麦转炉炼钢法和后来马丁平炉炼钢法的发明使生产建筑用钢的成本下降并且引发了建筑用钢在下个百年的快速发展。

钢最严重的缺点是容易被氧化，因而需要用油漆或一些其他涂料保护。当钢用于可能发生火灾的环境时，应该被包围在一些耐火的材料中，例如石料或混凝土。通常，钢的组合结构不易被压碎，除非是掺杂了冶金成分、低温以及空间压力存在的情况。

钢筋混凝土和预应力混凝土是主要的建筑材料。天然的水泥混凝土已经使用长达数世纪之久。尽管人造水泥被英国人 Aspidin 于 1825 年申请了专利，但现代混凝土建筑却兴起于十九世纪中叶。现代水泥是石灰石和黏土的混合物，两者被加热后碾成粉末。它可以在施工场地或靠近施工场地的地方和沙子、骨料（小石子、碎石或砾石）以及水混合制成混凝土。不同比例的成分可以制造出不同强度和重量的混凝土。虽然一些建筑者和工程师在十九世纪后期用钢筋混凝土做试验，但它作为一种占统治地位的建筑材料是在二十世纪初期。二十世纪后五十年钢筋混凝土结构设计和建筑得到迅速发展，早期即在法国的 Freyssinet 和比利时的 Magnel 被大量使用。

素混凝土作为建筑材料有一个非常严重的缺点：它的抗拉强度非常有限，只是它的抗压强度的十分之一。不仅素混凝土的受拉破坏是脆性破坏，而且受压破坏也是在没有多大变形预兆的情况下发生的准脆性破坏。（当然，在钢筋混凝土建筑中，可以得到适当的延性）。混凝土需要进行适当的养护和合理的选择并且掺入适当的混合添加剂，否则霜冻能严重损害混凝土。在长期荷载作用下混凝土在选择设计受压情况方面要仔细考虑。在硬化期和早期养护时，混凝土主要表现为收缩，因此需要添加适当比例的添加剂而且用适当的建筑技术来控制。

基于所有的这些可能的严重缺点，工程师已经试着为各种实际结构设计建立美观的、持久的和经济的钢筋混凝土结构。而正是因为设计尺寸和钢筋排列安排的谨慎选择和性比良好的水泥的发展，添加剂混合比例和混合配置的合理添加，加之养护技术和建筑方法、仪器的快速发展，工程师才能实现其优良作品。

混凝土具有多种用途，其组成材料广泛可取，并且能非常方便地浇制成满足强度及功能要求的形状。同时，随着新型预应力混凝土、预制混凝土以及普通混凝土施工方法令人兴奋的改善和发展潜力，这些因素综合起来使得混凝土在绝大多数结构中有着比其他材料更大的竞争力。预应力混凝土是钢筋混凝土的改良形式。钢筋弯成一定形状（弯成形）以给它们必要的抗拉强度等级。然后，通常通过先张法或者后张法，给混凝土施加预应力。预应力混凝土使得发展非常规形状的建筑成为可能，比如一些现代体育场，其巨大空间不被任何支柱阻断。

在现代，钢和加强钢筋在建筑结构中的使用量增加，而木材在建筑期间已经被撤到附属的、暂时的和次要的结构中使用，成为建筑材料的次要成员。然而，在二十世纪最后六十年中木材作为建筑材料又有恢复生机的迹象，改良的木材加工方法和各种不同的处理方法增加了木材的耐久性，而且叠片木材连同使用黏结技术的革命使得木材的性能有了更好的保证。各向同性的胶合板是最广泛使用的压层胶合板，随着技术的发展，压层胶合板已经发展成为特定的结构材料并对混凝土和钢造成了强大的冲击力。

将来可能发展的材料是工程塑料和稀有金属及其合金，如铍、钨、钽、钛、钼、铬、钒和铌。有许多不同的塑料可以用，而且这些材料所展现的力学性能在很大的范围内改变。因此，在具体设计应用方面为比较设计方案选择适当的材料是可能的。对塑料的使用受经验的限制。一般而言，塑料一定要与空气隔离，这是基于塑料结构在使用中的要求。塑料被应用的最有希望的潜能之一是嵌板和贝壳形结构。叠片或夹心嵌板已经被用于此种结构以鼓励未来建筑大量应用这个类型的材料。

另一种引起注意的是由纤维或像粒子的胶结钢筋的微粒组成的合成物材料，这种材料正在开发。虽然一种由玻璃或塑料胶结材料组成的玻璃纤维钢筋合成物已经使用长达数年之久，但是它们很可能衰落为次要的结构材料。钢筋混凝土是另一个积极地被研究而且发展的混合料。一些试验正应用在实际工作条件中，实验的主要内容为钢和玻璃纤维，但是大部分应用在钢纤维方面。

Unit 4 Bridges

Part One: Special Terms

1. 悬臂桥
2. 吊桥
3. 斜拉桥
4. beam bridge
5. truss bridge
6. arch bridge

Part Two: Situational Conversation

Exercises 2: Complete the Following Dialogue in English

(A: a student majoring in civil engineering   B: a teacher)

A: Can you tell me how to classify the buildings according to usage?

B: <u>Generally speaking, buildings can be classified into three general types according to their usage, namely the civil buildings, the industrial buildings and agricultural buildings.</u>（一般说来，建筑物按照使用情况可分为三大类，即：民用建筑、工业建筑和农业建筑。）

A: What are civil buildings?

B: <u>The civil buildings include the residential as well as the public buildings.</u>（民用建筑包括居住建筑和公共建筑。）

A: Can you list some examples of public buildings?

B: <u>Examples of public buildings are too numerous to list, such as building of offices, hotles, shops, schools, airports, stations.</u>（举不胜举。例如写字楼、酒店、商店、学校、机场、车站等都属于公共建筑。）

A: I see. Can I say the National Stadium is one of the most glaring examples of buildings?

B: <u>Of course you can, the National Stadium is not only spectacular but also strong, with the main structure designed for working life up to 100 years.</u>（当然可以，国家体育馆不仅壮观，而且坚固，其主体结构设计使用年限达到100年。）

A: Is there anything special in the design of its main structure?

B: <u>Yes, due to the need of a wider span, the roofs of the National Stadium are steel-structured, and the other load-bearing members use the reinforced concrete.</u>（是的，基于大跨度空间的需要，屋顶采用钢结构，其他主要承重构件采用钢筋混凝土结构。）

A: Is the National Stadium a high-rise building?

B: <u>Yes, for public buildings, those with more than 2 storeys over 24 meters high are high-rise buildings.</u>（是的，对于公共建筑而言，建筑物两层以上，高度24 m以上即为高层建筑。）

A: What about the high-rise for residential buildings?

B: <u>Residental buildings of more than 10 storeys can be regarded as the high-rise ones.</u>（十层以上的住宅建筑才算是高层建筑。）

A: Oh, I see.

B: **Part Three: Reading Comprehension**

1-5 DCBCA

Part Four: Translation Skills

Translation Exercise 1

The adoption of structural steel and reinforced concrete caused major changes in traditional construction practices. It was no longer necessary to use thick walls of stone or brick for multistorey buildings, and it became much simpler to build fire-resistant floors. Both these changes served to reduce the cost of construction. It also became possible to erect buildings with greater heights and longer spans.

Translation Exercise 2

The design and construction of buildings is regulated by municipal by laws called building codes. These exist to protect the public health and safety. Each city and town is free to write or adopt its own building code, and in that city or town, only that particular code has legal status. Because of the complexity of building code writing, cities in the United States generally base their building codes on one of three model codes: the Uniform Building Code, the Standard Building Code, or the Basic Building Code. These codes cover such things as use and occupancy requirements, fire requirements, heating and ventilating requirements, and structural design.

Unit 5　Building Code

Part One: Special Terms

1. 建筑部门
2. 下水道
3. 铸铁管
4. building code
5. workmanship
6. electrical system

Part Two: Situational Conversation

Exercise 2: Complete the Following Dialogue in English

(A: a supervisor　B: a welder)

A: <u>Did you have any training courses for safety education?</u>（你上过安全教育培训课吗？）

B: Yes. I learned all the safety regulations related to welding in the class.

A: <u>It's fine that you wear the helmet and protective goggles, but where are your hard shoes? Didn't your manager give one pair to you?</u>（你戴了安全帽和护目镜是对的，但

是你的硬底鞋在哪儿？你们经理没有给你一双吗？）

B: Yes, he did. I'm sorry. I forgot to wear them today.

A: <u>Don't forget again. Did you get a "Work Permit" for the welding here?</u>（别再忘了。你有焊接的工作许可证吗？）

B: Sure, here it is.

Part Three: Reading Comprehension

1-5 BDACD

Part Four: Translation Skills

Translation Exercise 1

A building code is a set of detailed regulations to ensure that all the buildings meet certain minimum standards of health and safety. Building codes have been enacted to protect citizens from any harm likely to come to them because of unhealthy or unsafe conditions.

Translation Exercise 2

The plan for all the new constructions must be approved by officials of the buildings departments before construction begins. They must be able to inspect all equipment, materials and workmanship before the building is approved for occupancy. If the equipment, workmanship, or materials do not meet the standards of the buildings, these officials have the right to order that the necessary changes be made.

参考译文

建筑规范

建筑规范是一套详细的条例，目的在于确保所有建筑物符合卫生和安全的标准。建筑规范的颁布，是为了保护其公民不会因不卫生或不安全的条件受到任何可能出现的危害。

所有新建工程的计划必须在施工开始前得到建筑部门的批准。他们必须在该建筑被批准占用前能够对所有设备、材料和工艺进行检查。如果设备、工艺或材料不符合建筑规范的标准，建筑部门有权命令进行必要的更改。

有时，业主可能想在电力、供暖或管道系统方面进行根本的变更，可能还想在结构方面也进行根本改变。在这种情况下，建筑部门必须事先批准修改。修改完成后，建筑部门必须再次检查工艺和材料。

起初，建筑规范是规定型的，它要求所有的建筑物都使用规定的材料，按规定的方法来施工。建造者在施工材料或施工方法方面几乎没有选择权。但是时代变了，新材料得以开发。第二次世界大战以来，性能类型的建筑法规有了很大的变化。在这种类型的规范中，材料或结构的性能标准得以突显。建造者能够自由选择符合这些标准的材料或建筑技术。

例如，对于一个住宅污水管道来说，规格型的建筑规范只会简单指定使用一定质量和尺寸的铸铁管。它还指定该管应按指定的方式安装。管道承包商在此方面没有选择权或发言权。但是，在一个性能型的建筑规范中，规范将详细说明管道不应受到管道掩埋处的污水或土壤中任何具有腐蚀性或有害性物质的影响。规范还要求该管道应当满足特定的最低强度要求。此外，规范将指出管道不应受到规定的范围内温度变化的影响。管道承包商可以自由选用塑料管、铸铁管，或者如果他愿意的话，还可以用黄金管，只要他能证明管道确实符合指定的标准。

Unit 6 Construction Cost Estimate

Part One: Special Terms

1. 被称做……

2. 源自于……

3. 与……平行

4. allocations of cost

5. budget cost

6. cash flow

Part Two: Situational Conversation

Exercise 2: Complete the following dialogue in English

A: What is the main purpose of control estimates?

B: <u>To monitor the project during the construction process, a control estimate is derived from available information to establish the budget estimate for financing, the budgeted cost after contracting but prior to construction, and the estimated cost of the completed parts in the construction process.</u>

（为了在施工期间对项目进行监督，根据可利用资料作出造价控制从而确定融资概算，签约后施工前的预算成本以及施工过程中完成部分的预算价值。）

A: Both the owner and the contractor must adopt some basic line for cost control during the construction, am I right?

B: <u>Yes. You are right. For the owner, a budget estimate must be adopted early enough for planning long term financing of the facility.</u>

（是的，对于业主而言，为了做项目的长期投资计划，必须尽可能地采用概算。）

A: Is it necessary for the owner to revise the budgeted cost?

B: <u>Yes. A revised estimated cost is necessary either because of the order changes</u>

initiated by the owner or due to the unexpected cost overruns or savings.

（是的，无论是由于业主变更订单或由于意外费用超支或节约，修正预算成本都是非常必要的。）

A: How about the control estimates for the contractor?

B: For the contractor, the bid estimate is usually regarded as the budget estimate, which will be used for controlling investment as well as for planning construction funds.

（对于承包商而言，投标报价通常被认为是概算，用于控制投资的目的以及规划建设资金。）

A: Should the budgeted cost be also updated periodically?

B: Yes. The purpose of updating the budgeted cost periodically is to reflect the completed estimated cost as well as to unsure adequate cash for the completion of the project.

（是的。概算定期修正的目的是反映完成的预算价值以及确保项目完成有足够的现金流转。）

Part Three: Reading Comprehension

1-5 ADCBC

Part Four: Translation Skills

Translation Exercise 1

Wall decoration shows the main part of the decorative works, thus easily leaving people first visual sense and impression, which is vitally important to the judgment of the decoration quality. Decorative works are to embellish the buildings, beautifully the environment and bring endless enjoyment to the mankind, so everyone decorator must have a high sense of responsibility and do his own work well.

Translation Exercise 2

If the cost of change orders are over RMB50,000 yuan, then all change orders issued because of the construction adjustments shall be submitted to Party A on the same day of the construction mid-term check and acceptance, and the approved amount will be paid to Party B with the construction mid-term payment; if the costs of the change orders are less than 50,000 yuan, then all change orders issued because of the construction adjustments shall be submitted to Party A on the same day of construction completion check and acceptance, and the approved amount will be paid to Party B with the construction completion payment.

参考译文

建筑造价估算

成本估算是项目管理中最重要的环节之一。它在项目建设的不同阶段为项目的成本建立了一条基准线。在项目开发过程中的某一特定阶段的成本估算就是造价工程师在现有数据基础上对未来成本的预测。根据美国造价工程师协会的定义，工程估价是运用科学理论和技术，根据工程师的判断和经验，解决成本估算、成本控制和盈利能力等问题的活动。实际上，所有的估价活动都是基于以下这些基本方法中的一种或几种的组合。

生产函数

在微观经济学中把过程的产出和资源的消耗这两者之间的关系叫作生产函数。在建筑工程中，生产函数则可认为是建设项目的规模和生产参数（如人工或资金）之间的关系。生产函数建立了产出的总量或规模与各种投入（比如人力、材料、设备）之间的关系。例如，代表产出的 Q 可以用代表各种投入的不同参数 $X_1, X_2, ..., X_n$ 等通过数学和/或统计方法表达。因此，对某一特定的生产，可以通过对各个投入参数赋予不同的值，从而找到一个最低的生产成本。房屋建筑的大小（用平方英尺表示）和消耗的人力（用小时/平方英尺表示）之间的关系就是生产函数的一个例子。

经验成本推论法

利用基于经验的成本函数估算成本需要一些统计技术，这些技术将建造或运营某设施与系统的一些重要特征或属性联系起来。数理统计推理的目的是为了找到最适合的参数值或者常数，用于在假定的成本函数中进行成本估算。在通常情况下，这需要利用回归分析法。

工程量清单单价

由工程量清单表达的各项任务或各个组成部分的单位成本能够明确，总成本就是各项产品的数量与其相应单位成本的乘积之和。单位成本法虽然在理论上非常直接，但是难以应用。第一步是将某工作分解成许多项任务，当然每项任务都是为项目建设服务的。一旦这些任务确定，并有了工作量的估算，用单价与每项任务的量相乘就可以得到每项任务的成本，从而得出每项工作的成本。当然，在不同的估算中对每项工作分解的详细程度可能会有很大差别。

联合费用的分摊

有时候要从现有的会计账目上去分解，从而确定某项具体操作的成本函数。这种方法的基本思想是，每一项花费都能对应地分配到操作过程中的某一特定步骤。理想情况是在成本分配过程中，混合成本能够有因果对应关系地被分解，并确定为某种基本成本。但是，在很多情况下，子项目和其分配成本之间难以确定或者根本不存在因果关系。例如，在建设项目中，基本成本可定义为以下5个方面：劳动力、材料、建筑设备、施工监理和普通办公管理费。然而这几个基本成本有可能会按比例地分配到工程子项的不同任务中去。

建造成本是整个项目成本中的一部分，虽然很重要，但只是施工项目经理控制下的一部分成本。在项目建设的不同阶段对估价精确度的要求也不一样，早期只是粗略估计，而到施工前的预算就应当相当可靠了。由于在项目生命周期早期设计方面的决策比后期更加具有不确定性，所以也就不能期望早期阶段的估价会准确。总的来说，估价的准确程度将反映估价时所获取的信息。

由于不同机构的要求不同，对建造成本估算也存在不同的观点。虽然在项目的不同阶段，建造成本的估算有许多不同的方法，但是根据其功用可分为三种主要方法。建造成本估算主要服务于以下三个方面：设计、投标和控制。对项目进行融资，要么需要设计估价，要么需要投标估价。

设计估价 对于业主或他指定的设计者而言，在规划和设计过程中要平行进行的估价的种类如下：

（1）匡算（宏观估价）；
（2）初步估算（概念方案阶段估价）；
（3）详细估算（明确估价）；
（4）工程师基于施工计划和说明的估算。

对于以上每一个估价阶段，设计所提供的信息量是逐步增加的。

投标报价 对于承包商来说，提交给业主的投标估价的目的是为满足竞争的需要或者是与业主谈判的需要，投标估价中包括直接施工成本（包括现场监督以及在此基础上增加一笔总体管理费）和利润。用于投标估价中的直接施工成本常用以下几种方法组合计算：

（1）根据分包商的报价；
（2）估计的工程量；
（3）施工方案。

造价控制 为了在施工期间对项目进行监督，根据可利用资源作出造价控制，从而确定融资概算、签约后施工前的预算成本、施工过程中完成部分的预算价值。

设计估算

在项目的规划和设计阶段，不同的设计估算反映了设计进展的不同阶段。在最初阶段，投资匡算或者宏观估价通常是在项目设计之前作出的，因此必须依靠过去类似项目的数据。初步估算或者称为概念方案阶段估价主要基于概念设计方案进行估算，此时只明确了最基本的技术方案。详细估算或者称为明确估价是在工程范围基本明确、详细设计正在进行、项目的基本特征已经确定时作出的。工程师基于施工计划和说明的估算要以明确的施工计划和施工说明为基础，并且业主也可以据此发包选择施工承包商。在做这些估算时，设计人员应将承包商的日程管理费和利润考虑进去。

为了进行成本估算，可以把一项设施的有关成本分解成几个适当的层次。

层次划分的详细程度取决于成本估算的种类，例如，对于概念设计方案阶段的估价，层次的划分就比较粗；而对于详细设计阶段的估价，层次的划分就应该很细。

例如，要对某条河上的拟建桥梁进行造价估算，投资匡算是对每个潜在的方案进行估算，如是采用拱桥方案还是悬臂桁架桥方案。当桥的类型确定后，比如选定拱形桥而不是其他形式的桥，就要基于拱形桥的平面布置，在初步或概念设计的基础上做初步估算。当详细设计进行到一定阶段，最根本的详细设计已经完成时，就要基于项目的明确范围进行详细估算。当完成了详细施工计划和施工指导说明后，工程师就可以根据各项工作和其工程量进行估算了。

投标报价

承包商的投标估价常常反映了承包商对完成投标工程的期望值，也反映了其所采用的估价工具。有些承包商拥有良好的估价程序，而有些则没有。由于在大多数工程招标中一般都是最低价中标，所以对于未中标者来说，在投标估价过程中的任何投入都是浪费。因此，如果一个承包商认为自己胜出的机会不是很大，他就会花费尽可能少的精力来进行成本估价。

如果一个总承包商在设施的建设过程中希望将工程进行分包，他有可能要求各个专业分包商提交分包报价单，所以总承包商将把成本估算的任务转移到分包商头上。如果所有的工程或部分工程由总承包商来施工，投标估计就要根据业主提出的计划和工程量或由承包商提供的施工组织计划来进行编制。本可以从某些商业出版物中找到其单价，再根据工程量就可以算出成本。但是，项目的实际情况可能导致实际施工程序与一般设计要求不同，在这种情况下承包商就要想办法估算其真实的成本和相关的费用。因此，为完成不同任务所需要的劳动力、材料和设备等就应该作为成本估算的参数。

造价控制

在施工过程中，业主和承包商都会有一个成本的基准来控制实际成本。对业主来说，必须尽早明确预算用于项目的长期财务规划。因而，在工程师估算完成前的相当长时间里，只能用详细估算来作为控制预算，因为详细估算也能充分明确地反映项目的范围。在工程进行的过程中，必须对成本估算进行定期更新，以反映完工前的预测成本。由于在施工中业主可能提出变更要求，或者有不可预见的突破成本或节约成本的情况发生，对成本估算进行及时更新是很必要的。

对承包商来说，经常用投标估价作为预算，既可以进行成本控制，也可以进行施工期间的财务规划。该预算成本也要定期更新，以反映完工前的预测成本，并确保项目完工前的现金投入。

Unit 7 Tender Documents and Contracts

Part One: Special Terms

1. 质量和安全方案

2. 工程量清单

3. 开工和竣工日期

4. certificate

5. tender document

6. Letter of Agreement

Part Two: Situational Conversation

Exercises 2: Complete the following dialogue in English

A: Nice to meet you!

B: Nice to meet you too! <u>Welcome to our company for tendering of the power station project. Here is the tender document for the works.</u>（欢迎来本公司参加发电厂项目的投标。这是本项目的投标书。）

A: There are so many volumes.

B: <u>Yes. Volume 1 is the Contract Conditions, Volume 2 is the Bill of Quantities, Volume 3 is Scope of Works, Volume 4 and Volume 5 are Technical Specifications.</u>（是的。卷1是合同条件，卷2是工程量清单，卷3是工程范围，卷4和卷5是技术规格书。）

A: There must be a lot of requirements for technical issues, otherwise these two volumes wouldn't be so thick.

B: <u>You are right. By the way, the engineers and supervisors working on this project should not only be able to read and use the technical specifications, but also remember that English is the contract working language.</u>（是这样。顺便说一下，参与本项目工作的工程师和监理不仅要能阅读和运用这些技术条件，还要记住英语是本合同的工作语言。）

The technical and supervisory staff, even your foremen for the site, should be able to speak English. Now let's go on to the Tender Documents.

A: We noticed that Volume 6 is Quality Assurance. <u>Would you give us some explanation about it?</u>（对此，你能向我们解释一下吗？）

B: Yes, of course.

Part Three: Reading comprehension

1-5 DDBAC

Part Four: Translation Skills

Translation Exercise 1

The Letter of Agreement is a promise of the contractor to the employer. It mentions the total initial contract price with the capital letters. The commencement date, the completion date and the maintenance period are also mentioned. This letter must be signed by the legal representative of the contractor or his authorized person.

Translation Exercise 2

Construction Method Statement consists of the following attachments: construction method description, construction schedules, site organization chart, key persons' information, workmen schedules, construction equipment schedules, layout for production area and living area, charts of electricity and water requirement, list of main supplier and subcontractors, list of repair parts, and so on.

参考译文

标书和合同

承包商应该先通过资格预审程序。按照文件要求，承包商应该在第一页填写公司的全称、法定代表人、公司地址、电话和传真号码、邮编和电子邮件地址（如果可用）。承包商还要填写其他表格。例如，对于员工的条件，他应该把高级技术人员、管理人员和熟练工人的数量写进去。至于建筑设备，要列出主要设备的数量、规格、折旧年限及厂家名称。如果承包商可以提供一份由正规大银行所开具的证书，则说明他的公司在该银行有具良好的信用记录，这将是非常有用的。

业主检查资格预审文件之后，该公司则可以进入下一阶段进行项目的投标。

招标文件包括四个部分。

第一部分：协议书。

第二部分：施工方案。

第三部分：质量和安全方案。

第四部分：工程量清单。

协议书是承包商对业主的承诺。该协议书用大写明确了初始合同的总价格，以及开工日期、竣工日期和保修期。这份协议书必须由承包商的法人代表或其授权人签名。

施工方案中包含以下附件：施工方案内容、施工进度、现场布置图、主要人员资质、工人们的安排、施工设备清单、生产区域和生活区域布局、电力和水需求布置、主要供应商和分包商的名单、备件清单等。

第三部分包括质量和安全方案。采取一定措施保证施工安全，并提交测试报告。承包商应根据要求准备好自己的安全方案。

最后一部分是最重要的一个。每个单价应包括材料成本、人工成本、机器成本和间接成本，其中包括员工工资、办公费用、住宿费用、工人交通费、培训费、利润、税金、银行担保费和保险费等。

Unit 8　Construction

Part One: Special Terms

1. 工作效率

2. 仓库

3. 碎石机

4. within the required time

5. at the lowest cost

6. constructor

Part Two: Situational Conversation

Exercise 2: Complete the Following Dialogue in English

A: <u>How many people is the project management composed of?</u>（该项目管理层由多少人组成？）

B: About 200. The top management has three persons only. They are Project Manager, Deputy Project Manager and Chief Engineer.

A: <u>How many departments are there under the project management?</u>（该项目管理下有多少个部门？）

B: Five. They are Works, Technology, Goods, Administration and Quality Departments.

A: <u>What about the function of each department?</u>（每个部门的职责是怎样的？）

B: The Works Department is responsible for all construction sections and terms. The Deputy Project Manager is also the manager of Works Department. The Technology Department is in charge of all matters about drawing and designing of temporary facilities. The Goods Department is in charge of material, construction equipment supply and the warehouse. The Administration Department is in charge of all matters related to finance and public relationship. The Quality Department is a special agency for it is controlled by not only Project Management but also the Headquarters of the company.

A: <u>Can you tell us something about the key persons for the project management?</u>（你能跟我们说说项目管理的重要组成成员吗？）

B: Sure. They are all qualified engineers with at least 10-year experience in the construction field. Here are their resumes including the proposed positions for the project, their education degrees, previous experience, especially the projects in which they have participated.

A: I want to remind you that as long as these key staff members are approved to work for the project, they should not be removed from the site without our permission. <u>By the way, I also want to know something about your foremen. (顺便说一下，我还想了解一下你们的工头。)

Part Three: Reading Comprehension

1-5 ACDAB

Part Four: Translation Skills

Translation Exercise 1

Construction is the translation of a design to reality. It is as important and complicated as the design in order that the structure will perform as it was intended to and the work is finished within the required time at the lowest cost. The designer must be in close contact with everything that is done during the construction work so that any changes in the site conditions, materials and work being done can be evaluated and, if necessary, corrected or improved.

Translation Exercise 2

The constructor should have the same knowledge of the working-plan as the designer. They must also know the details of the design and must understand any unusual aspect of the design. In fact, the constructor should go to the engineer for information and advice during the design stage so that the plans do not call for something that cannot be built economically. Both the designer and constructor must always work in harmony.

参考译文

施工

施工是把设计转化为实体。它与设计一样重要和复杂，以便设计内容能如预期一样变成现实，并且该工作要在规定时间内以最低的成本完成。

设计师必须密切接触施工工作中的一切事务，以便对现场条件、材料以及进行中的作业出现的任何更改进行评估，如有必要，还要进行修正或改进。

施工人员应当和设计师一样具有施工平面图方面的知识。他们还必须知道设计的细节，必须了解任何例外的设计情况。事实上，施工人员在设计阶段应该向工程师寻求信息和建议，以便设计上不致出现某些在经济上不能承受的项目。设计师和施工人员必须协调一致地开展工作。

更多的工人受聘于高峰时期。应当尽早给予工人工作技能和质量安全知识方面的培训。这将很大程度提高工作效率。

施工工作可分为以下几个阶段：

（1）方案、规范、现场基本需求和特性的评估；

（2）施工方案和进度；

(3）施工现场准备；

（4）主体施工；

（5）工完场清。

第一阶段的评估包括详细设计需求评估和现场需求评估。往往出现这种情况：第三阶段和第四阶段的工作正在进行，而第一阶段的评估还没有完成，此时就来不及了。

第二阶段是最重要的，如果要节约成本的话，施工中每个阶段的设备、人工和材料必须在准确的时间提供。

第三阶段包括铺通道，准备好仓库、切割机、混凝土搅拌机、办公室和工人住房。当然，这个工作往往只是开始，这些安排在施工过程中会多次改变。大部分时间和资金都花在施工阶段。

随着科技的发展，各领域的施工方法将会改变。许多日常施工将实现自动化和信息化，尤其是在住宅施工中。因此，施工阶段将需要更多的专业型人才，尤其是受过技术训练的人才。